Chemistry of Space Exploration

A Concise Introduction

Copyright © 2024 by Pestine Mards
All rights reserved. No part of this publication may be reproduced, distributed, or transmitted in any form or by any means, including photocopying, recording, or other electronic or mechanical methods, without the prior written permission of the publisher, except in the case of brief quotations embodied in critical reviews and certain other noncommercial uses permitted by copyright law.

Contents

1 **Introduction to the Chemistry of Space Exploration**   **11**
   1.1   Overview of Space Exploration and Its Importance ..   11
   1.2   Basic Concepts in Chemistry Relevant to Space   13
   1.3   Historical Milestones in Space Chemistry   15
   1.4   Interdisciplinary Nature of Space Chemistry   16
   1.5   Key Challenges in Space Chemistry   18
   1.6   Importance of Chemistry in Space Missions   20
   1.7   Chemical Analysis Techniques Used in Space   21
   1.8   Simulation and Modelling of Space Environments ..   23
   1.9   Case Studies: Chemistry in Famous Space Missions .   25
   1.10   Summary and Key Takeaways   27

2 **The Chemical Elements of Space**   **29**
   2.1   Composition of the Universe: Elemental Abundance .   29
   2.2   Hydrogen and Helium: The Primordial Elements ..   31
   2.3   Heavy Elements Synthesis: From Big Bang to Stellar Nucleosynthesis   33
   2.4   Distribution of Elements in our Solar System   35
   2.5   The Role of Carbon and Other Life-supporting Elements   36
   2.6   Rare and Precious Elements in Space: From Iron to Uranium   38
   2.7   Trace Elements in Different Space Environments ...   40

	2.8	Detecting Elements: Techniques and Tools in Astrophysics	42
	2.9	Implications of Elemental Composition on Space Exploration	43
	2.10	Future Research and Exploration Based on Elemental Studies	45
3	**Chemical Reactions in Space Environments**		**49**
	3.1	Understanding Chemical Reactions: Basics and Definitions	49
	3.2	Unique Characteristics of Space Environments	51
	3.3	Effects of Microgravity on Chemical Reactions	53
	3.4	Radiation-Induced Chemistry in Space	55
	3.5	Chemical Reactions in the Interstellar Medium	56
	3.6	The Role of Temperature and Pressure in Space Reactions	59
	3.7	Catalysis in Space: Natural and Engineered	60
	3.8	Chemistry on Planetary Surfaces: Mars, Venus, and Beyond	62
	3.9	Synthesis and Decomposition Reactions in Different Atmospheres	64
	3.10	Future Technologies for Controlling Chemical Reactions in Space	66
4	**Rocket Propulsion and Fuels**		**69**
	4.1	Introduction to Rocket Propulsion: Principles and History	69
	4.2	Chemical vs. Non-Chemical Propulsion: A Comparative Overview	71
	4.3	Common Chemical Rocket Fuels and Their Characteristics	73
	4.4	Chemistry of Combustion in Rocket Engines	74
	4.5	Liquid Fuels: Composition, Storage, and Handling	76
	4.6	Solid Rocket Fuels: Advantages and Composition	78

CONTENTS

- 4.7 Hybrid Rocket Fuels: Combining the Best of Both Worlds 80
- 4.8 Environmental Impact of Rocket Fuels 82
- 4.9 Innovations in Green Propulsion Technologies 84
- 4.10 Future Trends in Rocket Fuels and Propulsion Systems 86

5 The Role of Water and Ice in Space Chemistry — 89

- 5.1 Importance of Water and Ice in Space Exploration .. 89
- 5.2 Physical and Chemical Properties of Water and Ice in Space 91
- 5.3 Water and Ice in the Solar System: Distribution and Sources 93
- 5.4 Chemical Reactions Involving Water in Space 95
- 5.5 Ice in Asteroids and Comets: Composition and Significance 96
- 5.6 The Role of Water in Supporting Life on Other Planets 98
- 5.7 Techniques for Extracting and Analyzing Water and Ice in Space 99
- 5.8 Water as a Shielding Material against Space Radiation 101
- 5.9 Utilization of Water and Ice for Fuel and Life Support Systems 103
- 5.10 Challenges in the Management and Recycling of Water in Space Habitats 105

6 Atmospheres of Planets and Moons: Chemical Analysis — 109

- 6.1 Introduction to Planetary Atmospheres 109
- 6.2 Fundamentals of Atmospheric Chemistry 111
- 6.3 Common Gases in Various Planetary Atmospheres .. 113
- 6.4 Analytical Techniques for Atmospheric Analysis ... 114
- 6.5 Venus: Atmospheric Composition and Chemical Phenomena 116
- 6.6 Mars: Searching for Signs of Life through Atmospheric Chemistry 118

- 6.7 Giant Planets: The Role of Hydrogen and Helium in Atmospheric Chemistry 120
- 6.8 Titan and Other Moons: Complex Organic Chemistry in the Atmosphere 122
- 6.9 Impacts of Solar and Cosmic Radiation on Atmospheric Chemistry 124
- 6.10 Future Missions and Technologies for Atmospheric Exploration 126

7 Astrobiology: The Chemistry of Life beyond Earth 129
- 7.1 Introduction to Astrobiology: Scope and Significance 129
- 7.2 Chemical Building Blocks of Life in Space 131
- 7.3 Habitability Criteria for Life in Space 133
- 7.4 Biochemical Adaptations to Extreme Space Environments 135
- 7.5 Water and Organic Molecules: Key Ingredients for Life 136
- 7.6 Methanogenesis and Photosynthesis in Extraterrestrial Environments 138
- 7.7 Techniques for Detecting Biomarkers in Space 139
- 7.8 Case Studies: Potential Life-Bearing Bodies in Our Solar System 141
- 7.9 Challenges and Ethical Considerations in Astrobiological Research 143
- 7.10 Future Directions in the Search for Extraterrestrial Life 145

8 Materials and Chemicals for Space Habitats 147
- 8.1 Introduction to Materials and Chemistry in Space Habitats 147
- 8.2 Properties of Materials Required for Space Habitats . 149
- 8.3 Selection Criteria for Chemicals and Materials in Space 151
- 8.4 Influence of Space Environment on Material Degradation 153
- 8.5 Radiation Shielding Materials for Space Habitats ... 154

8.6	Self-Healing Materials for Long-Duration Missions .	156
8.7	Sustainable Production of Materials in Space	158
8.8	Chemical Processes for Air and Water Recycling . . .	160
8.9	Advanced Fabrication Techniques: 3D Printing in Space	161
8.10	Case Studies: Material Use in the ISS and Mars Habitats	163

9 Chemical Sensors and Instruments for Space Missions — 167

9.1	Overview of Chemical Sensing in Space Exploration .	167
9.2	Fundamental Principles of Chemical Sensors	169
9.3	Types of Chemical Sensors Used in Space Missions . .	171
9.4	Role of Spectroscopy in Chemical Analysis	172
9.5	Mass Spectrometry Instruments for Space Applications	174
9.6	Gas Chromatography for Analyzing Planetary Atmospheres	175
9.7	Developments in Nanochemical Sensors for Space . .	177
9.8	Integration of Sensors in Robotic and Human Missions	179
9.9	Challenges in Operating Sensors in Extreme Conditions	180
9.10	Future Innovations in Chemical Sensing for Space Exploration	182

10 Future Directions in Space Exploration Chemistry — 185

10.1	Challenges and Opportunities in Space Exploration Chemistry	185
10.2	Advanced Propulsion Systems and Fuel Chemistry .	187
10.3	Biochemical Solutions for Long-Duration Space Travel	189
10.4	Smart Materials for Adaptive Space Structures	192
10.5	Enhanced Analytical Techniques for Space Sample Analysis	194
10.6	Artificial Intelligence in Chemical Research and Monitoring	196
10.7	Space Pharmacology: Chemistry for Astronaut Health	198

10.8 Eco-friendly and Sustainable Space Exploration Practices 200

10.9 Interplanetary Trade and Utilization of Space Resources 202

10.10 Educational and Collaborative Projects in Space Chemistry 203

Preface

Welcome to *Chemistry of Space Exploration: A Concise Introduction*, a comprehensive guide designed to explore the intricate relationship between chemical science and space exploration activities. This book aims to provide a foundational framework for understanding how chemical principles apply in the diverse and challenging environments of space. It is crafted for readers who are starting their journey into the discipline of space chemistry as well as seasoned practitioners or enthusiasts looking to deepen their understanding of this fascinating field.

The content of this book is structured to facilitate a sequential learning experience. Starting from basic chemical and space concepts, it evolves through more complex topics such as the chemical makeup of planets, the biochemistry of potential extraterrestrial life, and the development of materials and fuels for space missions. Each chapter is meticulously structured to ensure clarity and ease of comprehension, with a focus on real-world applications and cutting-edge technologies that are pivotal in current and future space missions.

Intended for a broad audience, this book is suitable for students in the fields of chemistry, astronomy, and astrophysics; researchers and professionals working in space research; and anyone with a keen interest in how chemical science propels our exploratory capabilities beyond Earth. Additionally, it serves as an educational resource for instructors and educators who are involved in teaching the intriguing interplay between chemistry and the cosmos.

This book is not just a collection of scientific facts; it is a carefully curated resource that emphasizes the critical role of chemistry in overcoming the challenges faced during space exploration. From the propulsion systems that launch spacecraft into orbit to the life-support systems that make living in space possible, chemical science is at the forefront of innovation and discovery.

As you embark on reading *Chemistry of Space Exploration: A Concise Introduction*, expect to gain a comprehensive understanding of the chemical fundamentals that support space missions, the innovative materials engineering necessary for space habitats, the analysis of extraterrestrial samples, and the visionary chemical experiments that could pave the way for the future colonization of other planets.

It is our hope that this book will inspire you, equip you with knowledge, and enhance your appreciation for the role of chemistry in expanding the frontiers of human knowledge and capability in space. The journey through the pages that follow is not just educational—it is a testament to the limitless potentials when chemistry meets the vast expanse of space.

Chapter 1

Introduction to the Chemistry of Space Exploration

This chapter sets the stage for understanding the essential role that chemistry plays in the field of space exploration. It provides an overview of how fundamental chemical principles are vital for advancing our capabilities to explore, inhabit, and utilize extraterrestrial environments. It begins with a historical perspective on major achievements in space exploration influenced by chemistry, followed by discussing interdisciplinary approaches and technological innovations. The chapter also addresses the challenges faced by scientists in applying terrestrial chemistry concepts in space and outlines future perspectives in the ongoing synergy between chemical science and space exploration endeavors.

1.1 Overview of Space Exploration and Its Importance

Space exploration, a field characterized by the human quest to explore celestial bodies beyond Earth's atmosphere, serves a dual purpose. It acts as a vital platform for scientific advancement and as a beacon for inspiring human innovation and collaboration. The exploration of space is not merely an extension of human presence into unknown realms, but also a critical field where chemistry plays an indispensable role.

The significance of space exploration is multifaceted, encompassing scientific, technological, economic, and philosophical aspects. Scientifically, it enables the direct study of extraterrestrial environments, objects, and phenomena, which in turn enriches our understanding of the universe, including the origins of the solar system, the formation of celestial bodies, and the possibilities of life elsewhere. Chemically, understanding the composition and properties of alien planets, asteroids, and comets helps us to decode the complex processes that govern the universe. For instance, the analysis of moon rocks brought back during the Apollo missions revolutionized our understanding of lunar geology and the early history of the Earth.

Technologically, the challenges associated with space exploration spur innovation in materials science, fuel technology, propulsion systems, and robotic engineering. The development of new materials capable of withstanding extreme temperatures, radiation levels, and the vacuum of space is closely linked to advancements in chemical engineering and materials chemistry. An example can be seen in the creation of new alloys and composite materials, designed specifically for spacecraft and astronaut gear.

Economically, space exploration leads to the development of new industries and opportunities. The satellite industry, crucial for communication, weather forecasting, and navigation, is a direct product of space research and development. Additionally, the burgeoning interest in mining asteroids for resources such as water, which can be split into hydrogen and oxygen for fuel, and precious metals, opens new avenues for economic exploitation, driven by chemical analysis and processing techniques.

Philosophically, space exploration challenges human understanding of existence and our place in the universe. It fosters a global perspective that emphasizes cooperation among nations and cultures, aiming for peaceful and collective human advancement. The shared global interest in exploring space is a testament to its importance in fostering international collaboration and peace.

In summary, the ongoing quest to explore space is deeply woven with the fundamentals of chemical science. Each milestone achieved in space exploration has been built upon the foundation of chemical principles, from understanding the nature of celestial bodies and developing life support systems for astronauts to designing fuels and materials for spacecraft. The importance of space exploration thus extends well beyond mere curiosity or the desire to experiment—it is a comprehensive quest for knowledge, survival, and prosperity in the grand cosmic arena.

1.2 Basic Concepts in Chemistry Relevant to Space

A profound grasp of basic chemical concepts is fundamental to the science of space exploration, as these principles govern the behavior of materials and reactions under the unique conditions found beyond Earth's atmosphere. This section will elucidate these critical concepts, ranging from atomic structure and chemical bonds to the behavior of matter in different states and the kinetics of chemical reactions, contextualizing each within the domain of space exploration.

Atomic Structure and the Periodic Table: Understanding the atomic structure is pivotal in space chemistry. Elements in space, just as on Earth, consist of atoms that have protons, neutrons, and electrons. Protons and neutrons form the nucleus, with electrons orbiting in electron shells. The arrangement of elements in the Periodic Table aids in predicting properties like reactivity and bonding capabilities, which are vital in making materials for spacecraft and in analyzing extraterrestrial matter.

Chemical Bonds and Intermolecular Forces: The vastness of space holds a variety of conditions that significantly impact the types of chemical bonds and the intermolecular forces that can exist. Covalent, ionic, and metallic bonds form the primary types of chemical bonds. Covalent bonds involve the sharing of electron pairs between atoms, a type seen in the water molecules of cometary ice. Ionic bonds, seen in salt like lunar regolith, involve the electrostatic attraction between oppositely charged ions. In the metallic bonds, electrons flow freely among a lattice of metal ions, seen in the structure of asteroids and spacecraft materials.

Phases and Phase Transitions: Matter in space encounters extreme temperatures and pressures, from the near absolute zero of the deep cosmos to the intense heat and pressure of stellar environments. Understanding phases (solid, liquid, gas) and phase transitions (melting, freezing, evaporation, condensation) is crucial. For example, water detected on Mars exists in different phases under varying environmental conditions, crucial for future habitation and utilization.

Chemical Reactions and Stoichiometry: The capacity to predict and balance chemical reactions is critical in space exploration for generating life support systems, propulsion, and managing waste. Stoichiometry allows the calculation of relative quantities of reactants and products in chemical reactions, essential in the precise preparation of chemical reactions, whether it's for fuel synthesis on Martian surface or managing breathable air aboard spacecraft.

Reaction Kinetics and Equilibria: Reaction rates and equilibria give insights into how quickly a reaction will proceed and what quantities of reactants and products will be present at equilibrium. In the low-gravity conditions of space, reaction kinetics can alter, affecting rates of reactions and the time to reach equilibrium. Understanding these aspects is key for processes like the closed-loop life support systems that recycle water and air in space habitats.

Chemical Thermodynamics: The role of thermodynamics in space exploration cannot be overstated. It deals with the relations between heat, work, temperature, and energy in chemical processes. In the vacuum of space, thermal control becomes vital as temperatures can vary dramatically. The study of thermodynamics aids in designing efficient thermal protection systems and managing the energy resources of space missions, ensuring that instruments and astronauts can operate effectively.

Spectroscopy and Materials Analysis: Spectroscopy, the study of the interaction between matter and electromagnetic radiation, is a pillar in the chemical analysis of space. It helps in identifying the composition of celestial bodies remotely. Emission, absorption, and scattering spectroscopy are widely used techniques in missions, for instance, to determine the surface composition of planets and moons, or analyzing distant stars and galaxies.

Understanding these basic chemistry concepts is instrumental in every facet of space exploration. From crafting materials resilient to harsh space environments, to developing sustainable life support systems, and the study of extraterrestrial chemistry, the application of these principles is as boundless as space itself.

1.3 Historical Milestones in Space Chemistry

The voyage of chemistry in the realm of space exploration has been marked by a series of groundbreaking advancements that have not only expanded our understanding of the cosmos but have also revolutionized the chemical sciences. The chronicle of these developments provides a profound insight into the indispensable role that chemistry has played in deciphering and harnessing the mysteries of space.

One of the earliest and most notable milestones occurred during the Apollo missions by NASA in the late 1960s and early 1970s. The pivotal achievement of these missions was not merely the lunar landing but also the chemical analysis of Moon rocks. These analyses, conducted both on the lunar surface by astronauts and on Earth using samples brought back, fundamentally altered our understanding of the moon's composition. Techniques such as X-ray fluorescence spectroscopy and mass spectrometry were utilized to determine the elemental and isotopic compositions, revealing similarities to Earth's crust and providing evidence for prevailing theories about the moon's origin.

Following the Apollo missions, the Viking landers' journey to Mars in 1976 marked another significant chemical milestone. The Viking landers carried out the first martian surface in-situ analysis. These landers were equipped with a Gas Chromatograph Mass Spectrometer (GCMS) and conducted the Labeled Release experiment, which aimed to detect metabolic processes performed by potential Martian organisms. Though the results regarding Martian life remains inconclusive, the missions provided a wealth of data about the Martian soil's chemical composition, identifying substances like carbon dioxide, water, and organic compounds, thereby enriching our knowledge of Mars' potential for habitability.

The 1990s witnessed further advancements with the deployment of the Hubble Space Telescope, which, through spectral analysis, has identified chemical compositions of distant planets and stars, illustrating the diversity of chemical elements that exist beyond our solar system. This capability has provided insights into planetary formation theories and the processes that govern the universe at large.

In the 2000s, the Stardust mission exemplified the use of chemistry in space exploration by capturing and returning comet dust particles from comet Wild 2. The analysis of these particles on Earth has offered clues about the early solar system's conditions and the prebiotic chemistry that may have led to life on Earth. The mission utilized a special aerogel to trap the particles without altering their chemical properties, showcasing innovative materials science integrated with chemical analysis techniques.

The ongoing exploration of Titan, Saturn's largest moon, through the Cassini-Huygens mission, has also yielded significant chemical insights. Huygens probe landed on Titan in 2005 and found the presence of rivers and lakes of methane and ethane. Through spectroscopy and GCMS, it analyzed the atmospheric composition, significantly enhancing our understanding of complex organic chemistry in extraterrestrial environments.

These historical milestones in space chemistry not only demonstrate profound achievements in the field but also reflect the collaboration between various scientific disciplines, bringing together geologists, chemists, astronomers, and engineers. Each venture into space powered by chemical science paves the way for future explorations, pushing the boundaries of what is possible in the ongoing quest to discover our universe's secrets. As we continue to explore newer frontiers, the role of chemistry, with its ability to decipher the molecular and atomic nature of foreign celestial entities, remains a cornerstone in unlocking the cosmos.

1.4 Interdisciplinary Nature of Space Chemistry

Space chemistry, an inherently interdisciplinary field, integrates concepts from various scientific disciplines to address the unique challenges posed by the space environment. The complexity of space exploration demands a collaborative approach where chemistry interacts with physics, biology, materials science, and engineering to develop robust technologies and systems crucial for extraterrestrial exploration.

One of the primary interactions is between chemistry and physics, particularly in the study of materials under extreme conditions found in space. For instance, the behavior of propellants, which are crucial for spacecraft propulsion, depends significantly on their chemical properties and the physical conditions such as temperature and pressure in space. The knowledge of physical chemistry is vital for optimizing these materials to enhance their performance and reliability in space missions.

Moreover, astrobiology, a field at the intersection of biology and chemistry, plays a critical role in the search for life beyond Earth. It relies extensively on chemical principles to understand the potential biochemical pathways that might support life. Studies involving the analysis of meteorites for amino acids and other organic compounds are illustrative examples where organic chemistry and biochemistry converge to provide insights into the extraterrestrial organic synthesis and its implications for life in the cosmos.

Materials science, coupled with chemistry, is quintessential in the development of materials capable of withstanding the harsh conditions of space, such as high radiation levels and extreme temperatures. For instance, the design of spacecraft materials that can resist degradation by cosmic radiation involves understanding the radiation chemistry and the chemical stability of polymers and other composites.

Engineering, particularly chemical engineering, is also deeply intertwined with space chemistry. The design and operation of life support systems in spacecraft, which recycle water and air, involve chemical processes such as catalysis and adsorption. These systems are indispensable for long-duration space missions, where the spacecraft must support life without resupply from Earth.

The interdisciplinary nature of space chemistry is also evident in the development of analytical techniques used in space missions. Spectroscopy, mass spectrometry, and chromatography are employed to analyze the chemical composition of celestial bodies. These techniques, while rooted in chemistry, require the integration of mechanical and software engineering to operate effectively in the space environment.

Environmental science contributes to space chemistry through the study of planetary geology and atmospheres. Understanding the chemical interactions within the atmospheres of other planets, such as Mars, involves atmospheric chemistry, which is crucial for assessing their potential habitability and protecting future astronauts from atmospheric hazards.

The interdisciplinary nature of space chemistry is fundamental in addressing the multifarious challenges of space exploration. By bridging various scientific and engineering disciplines, space chemistry contributes to the advancement of knowledge and technology essential for exploring and understanding our universe. As space missions become more ambitious, the role of interdisciplinary approaches in space chemistry will undoubtedly expand, paving the way for new discoveries and innovations.

1.5 Key Challenges in Space Chemistry

Understanding the key challenges in space chemistry is essential for advancing our capabilities in space exploration and habitation. While the potential benefits of extraterrestrial chemistry are vast, several formidable barriers exist in the application and manipulation of chemical processes outside the Earth's environment. This section delves into the primary challenges that scientists and engineers face when applying chemical principles in space, highlighting the impact on mission design, safety, and success.

One significant challenge is the effect of microgravity on chemical reactions. On Earth, gravity influences many aspects of chemical processing, from fluid segregation to convection currents within mixtures. In the microgravity environment of space, these forces are greatly diminished, which can alter reaction rates, pathways, and outcomes. For instance, without gravity-induced convection, heat distribution within reacting mixtures becomes more uniform, potentially leading to different product yields. Moreover, the absence of sedimentation affects the stability of colloids and suspensions, crucial for materials science and pharmaceutical applications.

Another concern is the containment and management of volatile and hazardous materials in spacecraft. The closed environment of a spacecraft amplifies the risks associated with toxic or reactive substances. An accidental release of such materials could be catastrophic in the confined and controlled atmosphere of a space habitat. For example, handling liquid fuels, which are both toxic and highly reactive, requires advanced containment systems to prevent spills and ensure the safety of the astronauts and equipment.

Radiation in space further complicates chemical stability and the reliability of chemical processes. High-energy cosmic rays and solar radiation can initiate uncontrolled polymerizations, decompose complex molecules, and create radical species from otherwise stable com-

1.5. KEY CHALLENGES IN SPACE CHEMISTRY

pounds. Understanding and mitigating the effects of radiation on chemical systems is crucial for the long-term success of space missions, particularly when planning for human habitation on other planets or moons where atmospheric and magnetic protections are minimal.

Temperature extremes also pose a great challenge. The vast differences in temperature experienced in space can profoundly affect the rate and equilibrium state of chemical reactions. For example, the dark side of the moon can witness temperatures plunging to approximately -173 degrees Celsius, while the sunlit side might experience temperatures as high as 127 degrees Celsius. Such fluctuations can make it difficult to design chemical processes that are robust enough to function efficiently across such a broad range of conditions.

Furthermore, the limited availability of resources in space environments emphasizes the necessity for recycling and efficiency in chemical manufacturing processes. The development of closed-loop systems for water and air purification on spacecraft and lunar or Martian bases requires innovative chemical techniques to ensure that minimal waste is produced and that essential resources are constantly regenerated.

Lastly, the psychological and physiological changes in astronauts due to prolonged exposure to microgravity also influence how chemicals are handled and utilized. These changes could potentially alter the metabolism of drugs, affecting their efficacy and safety, which necessitates thorough studies and adaptable pharmaceutical strategies.

The deployment of chemistry in space environments encompasses a series of challenging hurdles that define the scope and limitations of extraterrestrial chemical applications. Addressing these challenges not only pushes the envelope of our scientific understanding but also ensures the feasibility and safety of prolonged human activities beyond Earth. Continuous research and innovation are essential in overcoming these barriers, paving the way for successful and sustainable space exploration. This understanding forms a cornerstone for subsequent sections about the practical use of chemistry in space exploration missions.

1.6 Importance of Chemistry in Space Missions

Chemistry plays an indispensable role in the success of space missions, influencing the design, execution, and outcome of extraterrestrial endeavors. This section explores how chemical science underpins various aspects of space travel, from life-support systems to fuel technology, and how it is pivotal in the development of materials suitable for space environments.

The formulation and management of rocket propellants are primarily chemical challenges. Traditional chemical rocket propellants, such as liquid hydrogen and liquid oxygen, are studied extensively to optimize efficiency and minimize risks. The chemistry of hypergolic fuels, which ignite on contact with each other, is also crucial for spacecraft maneuverability and control. Ongoing research in green propellant alternatives suggests a further scope where chemistry can contribute to safer, environmentally friendly solutions that comply with future space missions' stricter environmental protocols.

Life support systems on manned spacecraft and potential space habitats rely heavily on chemical processes. Advanced knowledge of chemical reactions enables the development of systems that can support life by providing breathable air and potable water through recycling methods. Techniques such as electrolysis of water to produce oxygen, and the use of chemical scrubbers to remove carbon dioxide, are fundamental in maintaining a habitable environment in closed systems, far removed from Earth's natural life-supporting cycles.

Material science, a field deeply rooted in chemistry, is essential for the development of durable, resilient materials that can withstand the extreme conditions of space, including high radiation levels, extreme temperatures, and mechanical stress. Chemical innovations in polymer stability, lightweight composite materials, and radiation-resistant coatings enhance the safety and longevity of space structures, from spacesuits to spacecraft exteriors.

Chemistry also contributes significantly to the field of astrochemistry, which explores the molecular nature and chemical processes occurring in space. Identified chemical signatures guide the detection of planets and assess their atmospheres, potentially earmarking them for future exploration or even habitation. By understanding the chemical makeup of celestial bodies, scientists can better predict and prepare for the conditions astronauts might face, or find signs of past or existing life.

Fuel cell technology represents another chemically grounded area critical to space missions. The operation of fuel cells, which convert fuel into electricity through a chemical reaction without combustion, provides an efficient and consistent power supply for space applications. Advancements in catalyst materials increase the efficiency and reduce the costs of these systems, highlighting the role of chemists in innovating more sustainable and effective energy solutions for space environments.

The immense void of space presents unique challenges, many of which are addressed through chemistry. Studying extraterrestrial geochemistry allows scientists to understand the planetary processes and compositions, contributing to broader scientific fields such as cosmology and planetary science, and aiding in the mining prospects of these bodies for future colonization efforts.

The role of chemistry in space missions is multifaceted and profound. Every aspect, from launch to daily operations in space environments, and considerations for future planetary habitats, relies fundamentally on chemical principles and innovations. The continued intersection of chemistry with space exploration not only enhances the viability of these exciting endeavors but also propels human understanding and capabilities beyond the confines of Earth.

1.7 Chemical Analysis Techniques Used in Space

The unique environment of outer space poses several challenges that demand specialized analytical techniques to effectively study and utilize extraterrestrial materials. The absence of gravity, extreme temperatures, and the presence of cosmic radiation require adaptations and innovations in analytical chemistry methods that are essential for successful space missions.

One prominent technique used in space exploration is mass spectrometry. It is particularly valuable because of its ability to characterize the composition of planetary atmospheres and surfaces. For instance, the Sample Analysis at Mars (SAM) instrument on the Mars Science Laboratory's Curiosity rover uses gas chromatography mass spectrometry (GC-MS) to analyze the atmosphere and rocks of Mars. By heating soil and rock samples, SAM releases gases which are then separated via gas chromatography and analyzed by mass spectrometry to identify various organic compounds and elements.

Another key technique utilized is spectroscopy. Various forms of spectroscopy, such as infrared (IR), ultraviolet (UV), and Raman, provide critical insights into the molecular and atomic makeup of cosmic bodies. The Cassini spacecraft applied UV spectroscopy to discover and investigate the upper atmosphere of Saturn and its rings, revealing the chemical interactions driven by solar ultraviolet irradiation. Moreover, Raman spectroscopy, as employed by the ExoMars rover, is designed to determine the mineralogy and detect potential organic compounds on Martian soil and rocks directly.

X-ray fluorescence (XRF) spectrometry is also a fundamental tool in extraterrestrial chemical analysis, largely due to its capability to perform non-destructive analyses and its easiness of miniaturization. The ChemCam instrument on the Curiosity rover utilizes a laser-induced breakdown spectroscopy (LIBS) approach, which is a type of XRF where a high-powered laser pulse is directed at a sample, vaporizing it and emitting light that can be analyzed to determine the sample's elemental composition.

Electrochemical methods have also found their place in space exploration. On the Huygens probe, part of the Cassini-Huygens mission to Saturn's moon Titan, electrochemical cells measured the concentration of sea salts and organic compounds in the atmosphere. This type of analysis is crucial for understanding the chemical dynamics and potential habitability of extraterrestrial environments.

One of the latest advancements in space-bound chemical analysis techniques includes the use of microfluidic devices. These devices integrate multiple laboratory functions on a single chip of only millimeters to a few square centimeters in size, making them ideal for space missions where compact and efficient tools are mandatory. These devices can be used to analyze chemical reactions, properties of substances, or microbial life in extraterrestrial water bodies.

Lastly, the Alpha Particle X-ray Spectrometer (APXS), another instrument aboard NASA's Curiosity rover, demonstrates the use of particle-induced X-ray emission (PIXE). This technique involves bombarding a sample with alpha particles and analyzing the resultant X-ray emissions to ascertain the sample's elemental composition. It provides fast and precise measurements that are pertinent in the resource-limited setting of space.

Each of these techniques represents a significant stride forward in our ability to conduct detailed and multifaceted chemical analysis in the extreme conditions of space. Through such technologies, scientists can investigate celestial bodies in unparalleled detail, advancing our understanding of the cosmos and paving the way for future exploratory missions. Using these sophisticated analytical tools, researchers continue to unlock the mysteries of space, revealing not just the composition of other worlds, but potentially uncovering signs of extraterrestrial life.

Analytical chemistry in space exploration is a dynamic field that continually evolves as new challenges arise and new technologies are developed. Through the persistent advancement of these techniques, the synergy between chemical science and space exploration will undoubtedly flourish, providing deeper insights into the universe's most fundamental aspects.

1.8 Simulation and Modelling of Space Environments

Simulation and modeling are central to the advancements in the field of space exploration, serving as essential tools to pre-test hypotheses and predictions in environments that are often unfeasible or too costly to replicate experimentally. These techniques provide a virtual platform to explore the diverse and extreme conditions found in space, paving the way for safer and more efficient missions.

In space exploration chemistry, simulation, and modeling help researchers and engineers to understand and predict the behavior of materials, chemical reactions, and processes under diverse extraterrestrial conditions. Firstly, molecular dynamics (MD) simulations offer insights into the atomic-level interactions over time, allowing scientists to study the structural stability and reactivity of materials in the vacuum of space, extreme temperatures, and under cosmic radiation exposure. This method has been used extensively to investigate the potential decomposition pathways of spacecraft materials and to design materials with enhanced resistance to degradation.

Computational fluid dynamics (CFD), another pivotal modeling tool, is used to analyze and predict fluid behavior in the microgravity conditions of space. This is crucial in designing life support systems where the management of liquid waste and the efficient delivery of gases, like oxygen, depend heavily on predictable fluid behavior. CFD simulations help in optimizing these systems to ensure they function effectively in the absence of gravity, where traditional Earth-based systems would fail.

Thermochemical modeling is additionally employed to simulate thermal processes such as heating and cooling, which are crucial for managing the temperatures of spacecraft and instruments. These models help predict how materials will react to the drastic temperature changes experienced during a space journey, such as thermal degradation or sublimation of volatile compounds. Thermochemical data is instrumental in designing thermal protection systems for spacecraft re-entering the Earth's atmosphere or landing on other celestial bodies.

Moreover, integrating quantum chemical calculations with statistical mechanics allows for simulating chemical reactions that might occur in space. This hybrid approach is particularly beneficial for understanding complex chemical systems and predicting new compounds that could be synthesized under space conditions. These studies are crucial for the development of in-situ resource utilization technologies, which aim to manufacture essential materials directly on other planets or moons using local resources.

Real-world application of these simulation tools can be exemplified by the case of Mars Rovers, where advanced modeling predicted the behavior of perchlorates in Martian soil and helped devise analytical methodologies for in-situ analysis. These simulations ensure that when actual sampling and analytical chemistry are conducted in space missions, the risks of malfunction or misinterpretation of results are significantly mitigated.

Highly detailed galactic chemical evolution models also enable astrophysicists and chemists to understand better the distribution of elements and compounds across the universe. These models simulate processes such as nucleosynthesis in stars and the resulting chemical enrichment of galaxies, contributing to theories about the origins of life-supporting molecules in space.

To better prepare for extraterrestrial exploratory missions, environmental chambers here on Earth, like NASA's Space Simulation Vacuum Chambers, often complement these digital simulations. Inside these chambers, physical samples are exposed to conditions akin to those found in space, such as extreme temperatures, vacuums, and varying radiation levels. When combined with robust computational models, these simulations provide data that can validate the computational predictions, enhancing the synergy between theoretical and practical approaches in space chemistry.

The interplay between simulation and modeling forms a cornerstone in the chemical study of space environments, guiding the development of new materials, predicting chemical behavior, and ensuring the success of missions via detailed pre-launch testing. The continuous evolution of computational methods and their applications in space chemistry not only reduce costs and increase the feasibility of missions but also deepen our understanding of the universe in ways previously thought to be inaccessible.

1.9 Case Studies: Chemistry in Famous Space Missions

In this section, we will explore real-world examples to elucidate the role of chemistry in several notable space missions. These case studies paint a vivid picture of how chemical science is crucial in overcoming the unique challenges posed by space exploration.

The Viking Mars Landers (1976): One of the seminal missions that heavily relied on chemistry to probe the possibility of life on Mars was NASA's Viking program. Each lander was equipped with a Gas Chromatograph-Mass Spectrometer (GC-MS) and a Gas Exchange experiment, designed to identify organic compounds and monitor gaseous exchanges with Martian soil samples. The chromatographic separation of Martian gases and subsequent mass spectrometric analysis allowed scientists to pursue signs of microbial life. Although no definitive biological signatures were found, the chemical insights gained were pivotal in understanding the Martian atmosphere and regolith.

Hubble Space Telescope (1990 - Present): While not a chemical experiment in the traditional sense, the Hubble Space Telescope has significantly contributed to chemical astronomy through its capacity to identify chemical compositions in celestial objects. Utilizing its spectrographs, Hubble has enabled astronomers to determine the elements present in distant stars, planets, and galaxies, through the analysis of light spectrums. These observations have crucial implications on our understanding of stellar evolution and the chemistry of the universe.

Galileo Probe to Jupiter (1995): Another landmark mission was the Galileo Probe's descent into Jupiter's atmosphere. Equipped with a variety of spectrometers, the probe conducted direct chemical analyses of the Jovian atmosphere. It measured compounds such as ammonia, methane, and water vapor while also detecting more complex organic molecules. The chemical composition data provided by Galileo has been essential in modeling Jupiter's atmospheric dynamics and studying its deep atmospheric layers.

Curiosity Rover (Mars Science Laboratory, 2012 - Present): The rover's onboard laboratory, SAM (Sample Analysis at Mars), incorporates a suite of instruments including a gas chromatograph, a mass spectrometer, and a tunable laser spectrometer. This powerful combination lets Curiosity conduct precise isotopic and chemical composition analysis of organic compounds in Martian soil and rocks. Notably, SAM's analyses have detected chlorinated organic molecules, contributing to debates about past life on Mars and informing models of present and future habitability.

Rosetta's Philae Lander on Comet 67P (2014): Achieving the first-ever landing on a comet, Philae's contributions to space chemistry include the COSAC and PTOLEMY instruments, which analyzed the chemical composition of the comet's surface. Initial findings showed a complex array of 16 organic compounds, four of which were detected for the first time on a comet. This rich chemical data provides insights into the primordial matter from which the solar system was formed.

Overall, these case studies demonstrate the broad scope of chemical analysis in space exploration missions. From interpreting the atmospheric data of distant planets and moons to directly sampling extraterrestrial soils and ices, chemistry remains a fundamental tool in the quest to understand the cosmos. The continual development and deployment of chemical instrumentation in space exploration not only deepen our knowledge of the universe but also refine our capabilities to conduct science in extreme environments, paving the way for future exploratory missions.

1.10 Summary and Key Takeaways

Through this chapter, we have embarked on a comprehensive exploration into the indispensable role that chemistry plays in the realm of space exploration. The journey commenced with an overarching view of the significance of space exploration and its implications for scientific, technological, and societal advancement. The discourse evolved to encapsulate the foundational chemical concepts that underpin and propel the field of space exploration, setting a robust platform to delve into the subsequent themes.

Historical revelations in space chemistry were examined, highlighting milestone achievements that underscore the progressive integration of chemical science into space exploration. These milestones not only represent technical feats but also pivot points in our understanding and capabilities. The interdisciplinary essence of space chemistry emerged vividly through discussions, portraying how chemistry amalgamates with other scientific disciplines to foster innovations that cater to challenging space environments.

One of the main emphases of this chapter was on the key challenges that confront the application of terrestrial chemistry principles in the extraterrestrial contexts of space. Factors such as altered gravitational conditions, extreme temperatures, and the vacuum of space necessitate rethinking and adaptation of chemical principles that are otherwise standard on Earth. Our exploration of these challenges set the stage for understanding how chemistry is intricately linked to various aspects of space missions, from life support systems to fuel compositions and materials engineering.

In addressing the operational and analytical dimensions, we reviewed several chemical analysis techniques utilized aboard spacecraft and during missions. These techniques are pivotal in conducting scientific experiments and in situ analyses of planetary surfaces and atmospheres, thereby contributing critically to our expanding knowledge of space environments.

The simulation and modeling of space environments have been another focal point, facilitating the prediction and preparation for the conditions that space missions face. This section intertwined with various case studies drawn from notable space missions. These practical examples served as illustrative bridges between theoretical chemistry and real-world application in space exploration contexts, providing palpable connections to the concepts discussed.

As we conclude, it is evident that the chemistry of space exploration is a dynamic and ever-expanding field, marked by continuous learning and adaptation. The key takeaways from this chapter underscore the adaptability of chemical science in overcoming the formidable barriers posed by space environments, and its crucial role in enhancing the safety, efficiency, and success of space missions. The insights garnered through this chapter should serve as both a foundation and an inspiration for future chemical studies and research directed towards space exploration endeavors.

As we propel forward, the interconnection between chemical science and space exploration will undoubtedly deepen, driven by mutual advancements and an unyielding quest for expanding human presence beyond Earth. This chapter lays down the fundamental aspects and invites continued exploration into this exciting and challenging intersection of chemistry and space science.

Chapter 2

The Chemical Elements of Space

This chapter delves into the distribution and characteristics of chemical elements found throughout the universe. It explores the origins of these elements from the Big Bang to stellar nucleosynthesis and their prevalence in various celestial bodies such as planets, moons, and asteroids. The discussion includes detailed analyses of how these elements are detected and measured, using advanced spectroscopic and remote sensing technologies, and examines their implications for understanding the cosmic landscape and guiding future space missions.

2.1 Composition of the Universe: Elemental Abundance

As we delve into the composition of the universe, it becomes essential to understand the elemental abundance and distribution that define cosmic chemistry. The universe, predominantly comprising hydrogen (about 74% by mass) and helium (about 24%), presents a fascinating array of minor yet crucially important heavier elements. These elements, though minor in concentration, play significant roles in the structural and evolutionary dynamics of celestial objects.

To quantify these abundances, astronomers rely on spectroscopy—a technique that analyzes the light emitted or absorbed by materials in space. The spectral lines, specific to each element, act as fingerprints that tell us not about the presence of an element, but about its concentration. This analysis reveals that elements like oxygen, carbon, neon, and nitrogen, while significantly less abundant than hydrogen or helium, are still common in the universe, making up roughly 1% of its mass.

The abundance of these heavier elements varies notably across different regions of the universe. In areas with recent or ongoing star formation, such as nebulae, the abundance of heavier elements can be higher due to the process of stellar nucleosynthesis. Stellar nucleosynthesis, the process by which stars produce new elements from nuclear fusion reactions, enriches the surrounding interstellar medium with heavier elements upon the star's death, particularly during supernova explosions. These cataclysmic events are key in the dispersal of heavy elements, such as iron, silicon, and nickel, into the broader cosmos.

The abundance of elements significantly influences the physicochemical properties of celestial bodies. For instance, planets and moons within our Solar System show a diverse range of elemental compositions, reflective of their varied origins and histories. Rocky planets like Earth have high concentrations of silicates and metals, whereas gas giants are predominantly composed of hydrogen and helium. This variation is also evident in meteorites, which provide direct samples of extraterrestrial material, often showing enriched abundances of rare metals like iridium and osmium.

This distribution of elements has crucial implications for space exploration and the search for life beyond Earth. For example, determining the elemental composition of Mars has been pivotal in assessing its potential for past or present life. Instruments like the Mars Rover's Alpha Particle X-ray Spectrometer (APXS) analyze soil and rock samples to determine their chemical composition, providing insights into the planet's geochemical history and habitability.

Furthermore, understanding the elemental abundance in space aligns with the broader goal of astromaterial science, which studies the material properties under extreme conditions found in space—high vacuum, radiation levels, and microgravity. This knowledge is pivotal not just for exploring potential habitable environments, but also for developing materials and technologies suited for future long-duration space missions.

Concerning theoretical and educational developments, the study of elemental abundance in the universe also enriches our understanding of the fundamental processes that govern chemical evolution. It provides a vital context for theoretical models that predict how different elements form under various cosmic conditions, thereby contributing to our broader understanding of the universe's evolution from the Big Bang to the present day.

Thus, the study of elemental abundance isn't merely about cataloging what exists; it's deeply intertwined with dynamic processes and phenomena throughout the universe, offering insights into the past and potential futures of celestial bodies, and guiding upcoming interstellar exploratory missions. As this field continues to evolve with advancements in observational technologies and theoretical frameworks, our comprehension of the cosmic chemical landscape will undoubtedly become richer and more intricate.

Indeed, the composition of the universe, with its vast and intricate patterns of elemental abundance, remains one of the most essential and continuously expanding areas of astrophysical research, holding keys to unlocking not only the secrets of our own Solar System but also the broader cosmic phenomena.

2.2 Hydrogen and Helium: The Primordial Elements

In the narrative of universal chemistry, hydrogen and helium occupy foundational roles as the most abundant and primordial elements in the cosmos. These elements are not merely prevalent; they are fundamental to the structure and evolution of the universe from its inception. Exploring their origin and significance offers profound insights into both astrophysics and chemical cosmology.

Hydrogen, represented by the symbol H, is the simplest and most abundant element in the universe, constituting about 75% of all baryonic (normal) matter by mass. This element is primarily found in its plasma state in stars, where it plays a critical role in stellar processes. Helium, displayed as He, accounts for about 23% of the universe's baryonic mass. Its production was predominantly during the Big Bang nucleosynthesis—an epoch that lasted about three minutes post-Big Bang, where temperatures and densities were favorable for the fusion of protons and neutrons into helium nuclei.

The predominance of hydrogen was initially established in the nascent universe, forming from the quark-gluon plasma as the universe cooled and expanded. Protons, or hydrogen nuclei, emerged as the universe underwent recombination, capturing electrons to form neutral hydrogen atoms. Concurrently, helium was synthesized from the coalescing of neutrons and protons in a series of nuclear reactions that significantly impacted subsequent chemical and stellar evolution in the cosmos.

Understanding these processes requires an evaluation of the key nuclear reactions, such as the proton-proton chain reaction prevalent in stars like our Sun. This series of fusion reactions converts hydrogen into helium, releasing vast amounts of energy that sustain stellar luminosity and heat. The reaction sequence can be simplified as follows:

$$4H \rightarrow He + 2e^+ + 2\nu_e + \gamma,$$

where H is hydrogen, He is helium, e^+ represents positrons, ν_e are electron neutrinos, and γ denotes gamma rays. This energy release is a cornerstone of stellar and galactic processes contributing to the thermal and radiative state of the universe.

In contrast to the interior of stars, the interstellar medium (ISM) provides a colder and less dense environment where hydrogen prevails mainly in atomic and molecular forms. Molecular hydrogen (H_2), although difficult to detect due to its nonpolar nature, is fundamental in the ISM, serving as the principal material from which new stars form. The presence of molecular clouds in space—dense regions rich in H_2—is where star formation is most active, highlighting the role of hydrogen in the continuing saga of stellar and galactic evolution.

Spectroscopic methods, such as absorption and emission spectroscopy, are employed to detect and analyze hydrogen and helium in space. Each element leaves a unique spectral fingerprint that can be observed as discrete lines in the spectra of stars and galaxies. For instance, the Lyman series and the Balmer series depict transitions in the energy levels of the electron in a hydrogen atom, visible in ultraviolet and visible regions, respectively.

These initial elements hold more than a spectral significance; their abundance and distribution influence the formation of celestial bodies and dictate the potential for more complex chemical formation, such as that of heavy elements in later-generation stars. Helium's role does not end within the Big Bang but extends as it serves in alpha processes within heavier stars leading to the production of elements like carbon and oxygen, critical for life as known on Earth.

Thus, hydrogen and helium are not just ubiquitous in the universe; they are also the linchpins in the developmental narrative of everything from the smallest atomic particles to the largest galactic structures. The study of these primordial elements is essential not just for understanding the cosmos' past but also for predicting its future progression and the potential for other life-supporting environments.

2.3 Heavy Elements Synthesis: From Big Bang to Stellar Nucleosynthesis

The synthesis of heavy elements through cosmic phenomena marks a cornerstone in understanding the chemical and physical evolution of the universe. Central to this discourse is the conceptualization of how elements heavier than hydrogen and helium are formed through processes initiated from the Big Bang to expansive stellar nucleosynthesis phases.

Starting with the Big Bang, which is predominantly recognized for the creation of hydrogen, helium, and trace amounts of lithium and beryllium, the environment lacked heavy elements which are integral to the later formation of galaxies, stars, and planetary systems. The Big Bang nucleosynthesis, occurring within the first three minutes of the universe's inception, set forth the primordial chemical abundance. Protons and neutrons began fusing under extreme temperatures, forming helium nuclei in a universe where temperatures exceeded 10 billion degrees Celsius.

As the universe expanded and cooled, these primary elements formed the first stars, leading to the onset of stellar nucleosynthesis. Stars function as cosmic furnaces where lighter elements undergo fusion to yield heavier elements through a collection of nuclear processes. The core conditions of stars, characterized by high temperatures and densities, facilitate the fusion of hydrogen atoms into helium. Further processes such as the carbon-nitrogen-oxygen (CNO) cycle enrich the stellar composition, contributing critically to the synthesis of elements up to iron.

Superior to the formation of iron, nucleosynthesis enters a phase driven by neutron capture processes, classified as the *s-process* and the *r-process*. The *s-process* occurs at relatively lower neutron densities and high temperatures, taking place predominantly in asymptotic giant branch (AGB) stars. This processed pathway is slower, with neutrons captured by nuclei over timescales allowing beta decay to compete, gradually building heavier elements up to bismuth.

Contrastingly, the *r-process* occurs in environments with extremely high neutron fluxes, such as those found during supernova explosions or the merging of neutron stars. The rapid absorption of neutrons before beta decay can occur leads to the formation of highly neutron-rich nuclei, which subsequently decay into stable heavier elements beyond iron, including uranium and thorium. This process is essential for explaining the presence of numerous heavy elements in the universe.

Recent advancements in astrophysical observational technologies have allowed for direct observation and analysis of these phenomena. For instance, the detection of the electromagnetic spectrum counterpart of gravitational waves from a neutron star merger has underscored the importance of *r-process* nucleosynthesis as a dominant source of heavy elements in the universe. It highlighted how mergers not only emit powerful gravitational waves but also seed the cosmos with elements that form the basis for everything from the iron in our blood to the gold and uranium that underpin critical industrial processes.

Through stellar nucleosynthesis, the universe gradually enriches itself, transforming from a simple primordial state into a complex assembly of systems featuring a vast array of elements. The sophistication of nuclear reactions in stars provides a comprehensive understanding of the temporal and spatial distribution of elements, shedding light on the chemical evolution of galaxies and the potential habitability of planetary bodies.

The synthesis of heavy elements from the Big Bang through various stellar processes not only enriches our universe but also enhances our understanding of the life cycle of stars and the overall dynamic architecture of the cosmos. Continued research and advancements in spectroscopic analysis remain pivotal in unraveling the further mysteries of stellar alchemy and its broader implications for cosmology and space exploration.

2.4 Distribution of Elements in our Solar System

The diverse distribution of chemical elements within our solar system provides a crucial context for understanding both its formation and potential for supporting life. This section focuses on how elements are spatially positioned across different celestial bodies including planets, moons, meteorites, and the interplanetary medium.

The solar system is primarily dominated by hydrogen and helium, which constitute approximately 98% of the total mass of the solar system. These elements are remnants from the initial formation of the sun, which is an average-sized star in the Milky Way galaxy. The sun's composition provides a vital clue to the solar system's primordial environment, which was heavily influenced by the solar nebula from which all other solar bodies were formed.

The terrestrial planets, comprising Mercury, Venus, Earth, and Mars, show a markedly different elemental makeup compared to the gas giants like Jupiter and Saturn. These inner planets are characterized by higher percentages of silicate minerals and metals such as iron, nickel, and magnesium. This composition suggests a high degree of differentiation, a process driven by internal heating which causes lighter materials to rise to the surface while allowing heavier materials to sink towards the core.

In contrast, the outer planets are mainly composed of gases and ices, with hydrogen and helium again playing a dominant role, along with methane, ammonia, and water ice. Their lower density indicates that these planets have retained much of the primordial matter from which they originated. This retention is attributed to their massive sizes, which gravitationally bound these lighter elements more effectively.

The role of meteorites as carriers of elemental information cannot be overstated. Chondritic meteorites, in particular, are believed to be the building blocks of the solar system. They are primarily made up of iron, nickel, and silicates, and their study has significantly contributed to our understanding of the solar nebula. Non-chondritic meteorites show evidence of processes such as melting and differentiation, pointing towards varied and complex histories resembling those of the terrestrial planets.

Furthermore, the distribution of elements in the Kuiper Belt and Oort

Cloud is indicative of the gradation seen as one moves away from the sun. The compositions of objects in these regions are chiefly determined by volatile substances frozen into ices, such as water, methane, and ammonia, which hints at the decreasing influence of solar radiation with distance from the sun.

To comprehensively analyze the distributional variation of elements, several advanced technologies and methodologies have been employed. Spectroscopic studies, both from ground-based observations and space missions, have been pivotal. They allow for the assessment of elemental compositions based on the light absorbed and emitted by atoms and molecules. Other techniques include mass spectrometry carried aboard robotic missions, which physically samples and measures the compositions of non-Earth bodies.

This elemental distribution has profound implications for space exploration and the potential for life beyond Earth. For instance, the presence of water ice on Mars and several moons of Jupiter and Saturn bolsters the hypothesis of life-sustaining environments outside Earth. The elemental diversity also provides insights into the processes that govern planet formation and the evolution of the solar system.

Problems for review:

- Calculate the expected spectral signatures of hydrogen and helium in the solar spectrum, taking into consideration their atomic weights and predominant isotopes.

- Analyze the difference in elemental composition between Earth's crust and the Martian surface, discussing potential geological processes that might explain these variations.

- Utilizing the data on elemental abundance in the interplanetary medium, design a hypothetical spectroscopic mission to measure these elements, articulating the types of instruments you would use and the scientific questions you aim to address.

2.5 The Role of Carbon and Other Life-supporting Elements

Carbon, often dubbed as the backbone of life, fundamentally anchors the intricate matrix of biological molecules in organisms across

2.5. THE ROLE OF CARBON AND OTHER LIFE-SUPPORTING ELEMENTS

Earth and potentially other celestial bodies. This section meticulously investigates the indispensable role of carbon and other vital elements such as hydrogen, nitrogen, oxygen, phosphorus, and sulfur—collectively known as CHNOPS—in sustaining life. Moreover, it extends the discussion to the distribution, interaction, and implications of these elements in extraterrestrial environments, which is crucial for future space explorations and understanding extraterrestrial life possibilities.

Carbon's unique ability to form four stable covalent bonds with other atoms—including itself—allows for the formation of complex, stable, and diverse organic molecules. Its tetrahedral bonding geometry enables the construction of various molecular structures, from simple hydrocarbons to intricate DNA and RNA molecules. The versatility of carbon-based compounds explains their predominance in known life forms and makes carbon an element of prime interest in astrochemistry and astrobiology.

Hydrogen, the most abundant element in the universe, plays a critical role in forming water (H_2O) and organic compounds. Water, vital for all known forms of life, acts as a solvent, medium for chemical reactions, and a temperature buffer. The presence of liquid water on a planetary body is one of the primary criteria in the search for life. For instance, the ongoing exploration of Mars and Europa, one of Jupiter's moons, focuses significantly on their capability to host water.

Nitrogen, another pivotal element, features prominently in the structure of amino acids, nucleic acids, and energy-transfer molecules like ATP. It is a fundamental component of the atmospheric makeup of many planets and is crucial in the formation of habitable ecosystems. Understanding the cycling of nitrogen, from atmospheric forms to usable biological forms, has implications not only for assessing the habitability of celestial bodies but also in designing life support systems for long-duration space missions.

Oxygen's role extends beyond its necessity for respiration in aerobic organisms. It is instrumental in the formation of silicates and oxides that constitute a major component of planetary crusts and is prevalent in water chemistry, influencing geological and atmospheric processes.

Phosphorus and sulfur, though less abundant, are equally critical. Phosphorus is a key element of ATP, phospholipids, and nucleic acids, acting as the energy currency within cells, while sulfur is vital in some amino acids and vitamins. These elements' geochemical cycles influence their availability for biological processes and their distribution and state are crucial for sustaining life.

Several spectroscopic techniques have been employed to detect and measure the existence and state of CHNOPS elements in various extraterrestrial environments. For example, infrared spectroscopy has detected water and carbon dioxide ices on Mars, the Moon, and several moons of the outer planets. Mass spectrometry aboard various space missions has allowed the direct analysis of the compositions of the atmospheres and surfaces of these celestial bodies.

Considering the implications of the elemental compositions, particularly the life-supporting elements, on space exploration, it is evident that their study not only aids in unraveling the cosmic origins and cycles of essential elements but also assists in selecting target planets and moons for future probes and manned missions. Understanding the distribution of these elements forms a cornerstone in astrobiology, planetary geology, and the burgeoning field of extraterrestrial resource utilization.

To conclusively fortify our comprehension of these essential elements and their cosmic interplay, ongoing and future missions should emphasize the development of more sensitive detection technologies, and strategies for in-situ resource utilization. This will critically support long-duration human space exploration and potential colonization efforts, ensuring sustainable access to life-supporting elements in the harsh environment of space.

2.6 Rare and Precious Elements in Space: From Iron to Uranium

As we continue our exploration into the chemical composition of the universe, it becomes imperative to focus on the rare and precious elements, particularly those ranging from iron to uranium. These elements not only play key roles in the physical structure and cosmic phenomena of celestial bodies but also hold profound implications for technological advancements and future space explorations.

The synthesis of elements heavier than helium, including iron and beyond—often termed as heavy elements—primarily occurs in the cores of massive stars through the process of stellar nucleosynthesis.

2.6. RARE AND PRECIOUS ELEMENTS IN SPACE: FROM IRON TO URANIUM

During the later stages of stellar evolution, as stars exhaust their hydrogen fuel, they undergo a series of fusion reactions forming heavier elements. Iron, which serves as a critical demarcation point in this process due to its peak binding energy per nucleon, thereby marks the limit of energy gain through fusion. The creation of elements heavier than iron—such as gold, silver, platinum, and uranium—requires a net input of energy and occurs predominantly during explosive stellar events such as supernovae and neutron star collisions. These cataclysms provide not only the necessary energy but also create an environment rich in neutrons, which rapidly capture into nuclei in a process known as the rapid neutron-capture process, or r-process.

Spectroscopic techniques are employed to determine the presence and abundance of these elements in space. By analyzing the spectra of stars and supernova remnants, astronomers can identify the characteristic absorption and emission lines of elements such as iron, gold, and uranium. This data is vital for understanding the distribution and concentration of these elements across the cosmos.

Iron, widely observed via its spectral lines in numerous stellar environments, is found to be a major component of the crustal material of rocky planets and moons, and in the metal-rich cores of such planetary bodies. The prevalence of iron plays a crucial role not only in the geological activity of these planets—including magnetic field generation and tectonic movements—but also in the potential habitability of planets.

Beyond iron, gold and platinum group elements, despite their rarity, have been detected in trace amounts on Earth's moon, Mars, and certain asteroids. The asteroid Psyche, largely believed to consist of a high concentration of metal with significant amounts of nickel-iron alloy, is an example of celestial bodies thought to be remnants of early solar system differentiation processes where heavy elements were segregated to the core of protoplanetary bodies.

Uranium, appearing in much smaller concentrations in the universe, is yet significant due to its radioactive properties, which leads to heat generation through radioactive decay—an essential element driving geological activity and surface renewal processes on rocky planets. Its presence in the Earth's crust, derived from supernova remnants, underscores its cosmological as well as geochemical importance.

Consequently, these heavy and precious elements—spanning from iron to uranium—are not only crucial in uncovering the mysteries associated with stellar deaths and the violent episodes of the cosmic theater but also in shaping the physical and chemical properties of planets and other celestial bodies. The studious examination of these elements provides astronomers and chemists with insights into past stellar events and potential guiding information for future comparative planetology studies and resource utilization in space.

The understanding of distribution and utilization of these elements continues to play an integrative role in the strategies devised for interstellar expeditions and resource planning in future space missions. By leveraging our knowledge of these rare and precious elements, humanity can pave the way for sustainable exploration and possibly colonization of other planets.

2.7 Trace Elements in Different Space Environments

Trace elements, though present in minuscule amounts, play pivotal roles in shaping the chemistry of various space environments. These elements, typically found at concentrations less than 100 parts per million, include such constituents as manganese (Mn), zinc (Zn), copper (Cu), and selenium (Se). Their distribution and impact across different celestial landscapes are vital in aiding our understanding of cosmic phenomena and the potential for life beyond Earth.

To comprehend the significance of trace elements in space environments, it is imperative to first consider their origins. Trace elements are primarily produced through stellar nucleosynthesis processes in the interior of stars and are expelled into space during supernova explosions. Subsequent dispersal of these elements into the interstellar medium enriches nearby forming celestial bodies, leading to their variable distribution across the universe.

In the context of planetary bodies, Mars serves as a prime example of trace element analysis. Spectroscopic data from Mars rovers and orbiters have detected varying levels of nickel (Ni), chromium (Cr), and bromine (Br), amongst others. These elements are crucial for understanding the planet's geological history and present conditions. For example, high concentrations of bromine in Martian soil are indicative of interactions with water and atmospheric processes, both of which are key considerations in the study of Mars' habitability.

2.7. TRACE ELEMENTS IN DIFFERENT SPACE ENVIRONMENTS

The role of trace elements is not limited to planets but extends to smaller celestial bodies such as asteroids. One illustrative instance is the study of the asteroid Bennu by NASA's OSIRIS-REx mission, which discovered bountiful amounts of cobalt (Co) and vanadium (V). These trace elements suggest a history of extensive heating and chemical processing, providing clues to the asteroid's origin and the early solar system's evolution.

Besides planetary bodies, trace elements influence the chemical makeup of interstellar clouds. Interstellar clouds rich in elements like iron (Fe) and magnesium (Mg) play a critical role in the formation of stars and planets. Furthermore, trace elements like iodine (I) and cadmium (Cd) have been identified in such environments through radioastronomy, revealing complex chemistry that could influence the synthesis of prebiotic molecules.

The study of trace elements also extends to exoplanets. Techniques like transmission spectroscopy have allowed for the detection of potassium (K) and sodium (Na) in the atmospheres of hot Jupiters. These findings are integral to modeling atmospheric processes and assessing the potential for chemical disequilibrium, which is a marker of potential biological activity.

The distribution and effects of trace elements across different space environments can be summarized with the aid of diagrams such as distribution graphs and element abundance charts. These visual aids help clarify the spatial variability of trace elements and their correlation with specific cosmic phenomena, enhancing comprehension and retention of these complex interactions.

While trace elements constitute only a tiny fraction of the elemental composition in the cosmos, their impact is profound. Continuing to explore and understand these elements in varied space environments not only enriches our knowledge of the cosmos but also aids in the ongoing quest for extraterrestrial life. The challenges faced in detecting and studying these elements push the boundaries of technology and exploration, paving the way for future innovations in space science.

To further solidify this understanding, students are encouraged to tackle the following problems:

1. Calculate the expected spectral lines for zinc when exposed to high-energy cosmic rays, using the provided spectroscopic data.

2. Analyze the correlation between trace element concentration and planet habitability using statistical methods from the data sets collected from Mars and Earth.

3. Develop a hypothetical model to predict trace element distribution in a newly discovered exoplanet atmosphere, considering stellar type and proximity to the parent star.

These exercises will enhance analytical skills and deepen the understanding of the integral role trace elements play in the broader context of space chemistry.

2.8 Detecting Elements: Techniques and Tools in Astrophysics

Astronomical spectroscopy stands as the cornerstone in the detection and analysis of chemical elements through various astrophysical media. This highly indispensable technique relies on the analysis of the light spectrum emitted, absorbed, or reflected by astronomical objects to determine their composition, temperature, density, and motion.

The basis of spectroscopy resides in understanding the atomic and molecular structure of elements. Each element possesses a unique spectral fingerprint—specific wavelengths at which they emit or absorb electromagnetic radiation. When light from celestial bodies passes through a prism or a diffraction grating in a spectrometer, it is dispersed into its component colors, producing spectra. By examining these spectral lines, astrophysicists can infer the presence and abundance of elements within stars, nebulae, and other celestial objects.

There are primarily three types of astronomical spectroscopy: emission, absorption, and reflection spectroscopy. Emission spectroscopy is used to study hot gases emitting light at characteristic wavelengths. Absorption spectroscopy, on the other hand, analyzes the light absorbed by cooler gases lying between the observer and a light source. Reflection spectroscopy examines light reflected off the surface of planets and moons. These spectroscopic methods have unveiled not only the presence of hydrogen and helium, the most abundant elements in the universe, but also heavier elements like iron, oxygen, and silicon.

Remote sensing techniques also play a pivotal role in the detection and quantification of elements in space exploration. Spacecraft equipped with spectrometers and sensors orbit celestial bodies, gathering data on their surface composition. One such example is the use of Gamma-ray and Neutron Spectrometers, which have been instrumental in determining the surface compositions of bodies like the Moon and Mars. These instruments measure the energy spectra of gamma rays and neutrons emitted due to the interaction of cosmic rays with the surface matter, providing insights into the presence of elements such as hydrogen, calcium, and iron.

Mass spectrometry is another vital tool, often used on interplanetary missions to analyze the composition of the atmospheres and surfaces. The device ionizes samples collected in situ or remotely, sorting the ions based on their mass-to-charge ratios. This technique was notably showcased by the Cassini mission, which utilized the Cosmic Dust Analyzer to examine the ice and dust particles in Saturn's rings and moons, revealing complex organic compounds and water ice.

The advancement in telescopic technologies has boosted the capabilities of these techniques. The development of ultra-sensitive detectors and higher resolution spectrometers dramatically enhances the accuracy and detail in which these analyses can be performed. Additionally, the integration of computational models and software has improved the simulation and interpretation of astrophysical phenomena, enabling a deeper understanding of the cosmic distribution of elements.

In synthesis, the dynamic field of astrophysics utilizes a sophisticated array of technologies and methodologies to explore and understand the elemental composition of the universe. Through the lens of spectroscopy, remote sensing, and mass spectrometry, coupled with advanced computational tools, modern astronomy continues to unravel the mysteries of the cosmos, paving the way for future explorations and discoveries.

2.9 Implications of Elemental Composition on Space Exploration

Understanding the elemental composition of celestial bodies not only enriches the field of astrophysics but also significantly influences the planning, execution, and long-term goals of space exploration missions. This foundational knowledge plays a critical role in designing spacecraft materials, determining fuel types, assessing potential colonization resources, and even selecting mission targets.

Materials engineering for spacecraft employs information about elemental distribution in space to ensure resilience and functionality under extreme conditions. For instance, the presence of titanium on the Moon and Mars directs research and development towards titanium-based alloys for construction due to their high strength-to-density ratio, resistance to space corrosion, and thermal properties. In addition, knowing the elemental makeup of asteroids rich in metals like nickel and iron has spurred the development of potential space mining missions that aim to harvest these metals for space-based manufacturing, reducing the need for Earth-based materials and thus minimizing launch weights and costs.

Fuel choices for space missions are deeply influenced by available resources. Hydrogen, found abundantly as ice on lunar poles or Martian subsurface, becomes a prime candidate for fuel via water electrolysis. This not only ensures a sustainable energy source but also reduces dependency on Earth's resources, paving the way for longer and potentially more distant missions. Furthermore, the discovery of helium-3 in lunar regolith—potentially a cleaner, more potent source of nuclear fusion energy—has implications for providing almost unlimited energy for future human habitats or spacecraft.

Planetary colonization and terraforming concepts rely heavily on a detailed understanding of a planet's elemental composition. Mars, for instance, with its deposits of carbon, nitrogen, and oxygen in the form of carbon dioxide and permafrost, could theoretically support plant life under controlled conditions, aiding in human colonization efforts. These elements are essential for biological life and could be manipulated to create Earth-like living conditions on Mars, making it a key candidate for colonization.

Mission targeting, an essential aspect of space exploration, is directly affected by the elemental mapping of celestial bodies. Scientific missions often target moons or asteroids with rare elements or peculiar chemical signatures that could unravel new physics or chemistry, potentially leading to new advancements in technology or theories in planetary science. For example, NASA's OSIRIS-REx mission to the asteroid Bennu involves understanding the carbon-rich material present on its surface, which is thought to be a remnant of the early solar system and could provide insights into the origin of life on Earth.

Moreover, safety protocols in space missions also leverage elemental data. The understanding of solar particle events and the elemental composition of solar winds helps in developing better shielding technologies for astronauts, protecting them from harmful radiation. Elements like boron, lead, and hydrogen, known for their radiation absorption capabilities, are studied to enhance the efficacy of protective gear and spacecraft shielding.

The strategic exploitation of knowledge about elemental constituents in space exploration holds the key to not only ensuring the feasibility and longevity of interplanetary travel but also fortifies the prospect of sustainable extraterrestrial colonization. This continuous loop of discovery and application propels both our capacity and understanding in the boundless journey through space, securing a foothold for humanity in the cosmic arena. This rich interplay between elemental science and practical application underscores the fundamental importance of chemistry in propelling and safeguarding the future of human space exploration.

2.10 Future Research and Exploration Based on Elemental Studies

Scientific inquiry into the elemental composition of celestial bodies not only enhances our understanding of the cosmos but also fundamentally guides the trajectory of future space missions and research agendas. As we advance our knowledge of chemical elements in space, the impact on exploration strategy and technology development becomes increasingly significant.

The elemental make-up determined via spectroscopic studies and remote sensing essentially informs the selection of targets for exploration missions. Bodies such as asteroids, moons, and planets that show an abundance of rare or potentially useful elements become priorities not only for basic scientific inquiry but also for their potential in-situ resource utilization (ISRU) capabilities. For instance, water ice discovered on the Moon and Mars can be split into hydrogen and oxygen, crucial for life support and as rocket propellant, respectively. Similarly, metals like iron found in the Martian soil could be used in the manufacturing of habitats and tools, reducing the need to transport materials from Earth.

Looking forward, ongoing and future research is expected to focus on refining our understanding of the distribution and concentration of these vital resources. Enhanced geologic mapping of celestial bodies using improved spectroscopic technologies will allow for more precise identifications of material compositions. These detailed maps will not only dictate the specifics of mission planning but could also lead to the discovery of new elements or minerals that have not yet been observed.

Another promising area of research is the study of elemental behavior in the unique conditions of space environments. Microgravity, high radiation levels, extreme temperatures, and other factors could alter the chemical properties and states of elements in ways not observable on Earth. Such studies could open up new avenues in materials science, with potential applications in spacecraft design, construction, and even terrestrial technology improvements.

Investigations into the origins of elements through processes like stellar nucleosynthesis continue to hold profound importance in theoretical physics and cosmology. Understanding the lifecycle and distribution patterns of elements post-supernova provides insights into the formation of stars and planetary systems. This can assist in predicting the types of elements likely to be found in undiscovered celestial bodies.

Machine learning and artificial intelligence are poised to play increasingly pivotal roles in the analysis of complex data collected from spectroscopy and other remote sensing technologies. These tools can help decipher patterns and correlations in elemental data, leading to quicker and more accurate predictions of elemental locations. Such capabilities will enable more targeted and efficient planning of exploration missions.

Finally, the ethical and practical implications of exploiting space-based resources must be considered thoroughly. As nations and private entities look to the cosmos for resources, international agreements and guidelines will be crucial in ensuring responsible and equitable use of these off-earth materials. Researchers will need to work closely with policymakers to navigate the challenges posed by such endeavors.

Advances in our understanding and technological abilities to detect and analyze elemental compositions in space are moving at an unprecedented pace. Future research will likely continue to unearth new knowledge that will prompt further questions about our universe while simultaneously providing the tools needed for deeper exploration and practical utilization of space-based resources. Through a combination of enhanced analytical tools, interdisciplinary collaboration, and thoughtful consideration of ethical implications, the study of space elements will remain at the forefront of both academic inquiry and exploration strategy for decades to come.

Problems for Reflection and Practice

- Given the potential of ISRU on the Moon, detail a hypothetical mission plan that utilizes local resources for establishing a sustainable human presence. Consider technological, logistical, and environmental challenges.

- Analyze the potential changes in element behavior under conditions encountered in space. Propose an experiment that could test your hypothesis aboard the International Space Station.

- Discuss the role of AI in analyzing spectroscopic data from telescopic observations. Design a simulated workflow incorporating machine learning for detecting rare elements on asteroids.

- Develop a policy proposal outline for the ethical extraction and use of elements from a newly discovered asteroid with high concentrations of platinum group metals.

- Reflect on the long-term implications of elemental studies in space on Earth's economy. Predict and debate foreseeable economic shifts or impacts.

Chapter 3

Chemical Reactions in Space Environments

This chapter explores the complexities and unique characteristics of chemical reactions occurring within various space environments, from the microgravity of space stations to the extreme conditions of planetary atmospheres. It covers how factors such as radiation, vacuum, and temperature extremes impact chemical dynamics differently than on Earth. The text examines specific reaction mechanisms, such as those driving the formation of complex organic molecules in interstellar clouds, and discusses the methodologies employed to study these phenomena, highlighting the adaptability and innovative approaches needed in extraterrestrial chemical research.

3.1 Understanding Chemical Reactions: Basics and Definitions

Chemical reactions are fundamental processes by which molecules interact to form new products, accompanied by the rearrangement of atoms and changes in their bonding structures. These reactions are governed by laws that are invariant across the cosmos, yet the context in which they occur can dramatically alter their paths and outcomes.

A chemical reaction can be generally described by a chemical equation, typically comprising reactants (the starting substances) and products (the substances formed as a result of the reaction). Each equation is balanced, meaning the number of atoms for each element is conserved, adhering to the Law of Conservation of Mass. For instance, the reaction between hydrogen and oxygen to form water is represented by the equation

$$2H_2 + O_2 \rightarrow 2H_2O.$$

Reactants are transformed into products through pathways that may involve various intermediate states or transition states, which represent high-energy configurations. These states are crucial for understanding the reaction kinetics, which describe the rate at which a reaction proceeds. Reaction rates can be influenced by several factors including temperature, pressure, the presence of catalysts, and the physical state of the reactants.

The concept of chemical equilibrium is essential when discussing reactions that can proceed in both forward and reverse directions. At equilibrium, the rates of the forward and reverse reactions are equal, leading to a constant concentration of reactants and products. The position of equilibrium is described quantitatively by the equilibrium constant, K, a unitless number providing a measure of the extent of a reaction at a given temperature.

Enthalpy (ΔH) and entropy (ΔS), two important thermodynamic parameters, also play vital roles in determining the spontaneity of a reaction. The Gibbs free energy (ΔG), derived from these parameters, is crucial for predicting the feasibility of reactions. A negative ΔG indicates a spontaneous reaction under constant pressure and temperature. The equation

$$\Delta G = \Delta H - T\Delta S$$

integrates these concepts, linking the heat exchange, disorder, and energy dispersal within a system.

In the context of space environments, these principles must be reconsidered and often reinterpreted. For instance, in microgravity environments, the lack of buoyancy-driven convection alters how substances mix and react, often leading to the enhanced importance of diffusion-controlled processes. Additionally, the extreme temperatures and radiation levels encountered in space can shift reaction equilibria and modify reaction rates.

Understanding these fundamental aspects provides a framework for discussing more complex chemical reactions and their unique characteristics in varied space environments as we proceed through this chapter. It sets the stage for a thorough exploration of how chemical reactions are tailored and observed in extraterrestrial settings, contributing to our broader comprehension of chemistry beyond Earth.

3.2 Unique Characteristics of Space Environments

Space environments present a plethora of unique challenges and characteristics that substantially influence chemical reactions. Unlike terrestrial conditions, space environments—encompassing everything from the near-vacuum of interstellar regions to the varied atmospheres of planetary bodies—exhibit extreme variations in pressure, temperature, and levels of radiative exposure which all play crucial roles in chemical dynamics.

The vacuum of space, characterized by extremely low pressures, is one of the most defining aspects of these environments. Terrestrial chemical reactions typically occur at pressures of one atmosphere, where molecules readily collide and react. In contrast, the near-absence of pressure in space means that molecules are far less likely to interact, significantly slowing the rate of many types of chemical reactions. However, this same characteristic can aid certain processes, such as the sublimation of ices to gases, which occurs more readily in vacuum conditions, thereby facilitating different types of reactions.

Temperature in space also exhibits extreme ranges which greatly influence chemical reactivity. In the shadowed lunar craters or on distant comets, temperatures can plummet to mere tens of degrees above absolute zero, significantly reducing the kinetic energy available for substances to react. Conversely, the unshielded solar radiation can heat exposed materials to several hundred degrees Celsius, potentially driving thermochemical reactions that would be non-viable under normal Earth conditions.

Radiation represents another pivotal factor in space chemistry. High-energy particles, such as those found in cosmic rays or solar wind, can interact with chemical substances, breaking chemical bonds and generating highly reactive ions or radicals. These particles, unlike anything encountered naturally on Earth, drive a broad array of reactions, including the polymerization of complex organic molecules and the cleavage of water to release hydrogen and oxygen. The ubiquitous presence of radiation in space not only initiates these processes but also alters the chemical pathways and mechanisms distinctly from those observed on Earth.

Moreover, space environments lack an atmosphere that can buffer or mediate the effects of radiation and temperature fluctuations. On planetary bodies devoid of significant atmospheric cover—like Mercury or the Moon—surface-bound chemicals experience a direct onslaught of solar radiation, leading to photodissociation or other light-driven reactions without the mitigating influence of atmospheric gases.

The microgravity conditions found in many space environments further alter chemical behaviors. On Earth, gravity affects fluid dynamics, sedimentation, and convection in ways that uniformly mix reactants and remove reaction products from the site of reaction. In microgravity, these processes are starkly different. Fluids and gases do not settle or layer in the usual way, which can lead to more homogeneous mixing or, conversely, the accumulation of reactants or products in unexpected regions, dramatically influencing reaction rates and products.

Understanding these unique aspects of space environments is crucial for the development of effective chemical management strategies in space exploration and habitation. Conditions that vary so markedly from those on Earth demand innovative approaches to chemical synthesis, containment, and control. Research in this area not only expands our fundamental understanding of chemistry under extreme conditions but also equips us with the knowledge to engineer materials and systems robust enough to handle the rigors of space travel and extraterrestrial colonization.

Recognizing and addressing these unique parameters provide the foundation for all further discussions on the specifics of space-bound chemical reactions, ensuring that science progresses safely and efficiently as humanity reaches into the cosmos.

3.3 Effects of Microgravity on Chemical Reactions

Microgravity, a condition encountered in space environments where gravitational forces are significantly less than those on Earth, presents unique challenges and opportunities for chemical research. The reduced gravity conditions found in spacecraft and space stations affect chemical reaction mechanisms, reaction rates, and the physical behavior of molecules and reactants. This section delves into how these conditions alter chemical kinetics and reaction equilibria, illustrating the necessity of reconceiving traditional chemical concepts when applied outside Earth's gravitational pull.

In microgravity, the absence of buoyancy-driven convection significantly impacts fluid dynamics. On Earth, convection currents are a primary mode by which heat and mass are transported within fluids. These currents are generated by the variation in density of fluids due to temperature or concentration gradients, which are themselves gravity-dependent. In the microgravity environment of space, these gradients do not result in convection, leading to a more uniform distribution of molecules in a solution. This homogeneity can alter rate constants and shift equilibrium states in chemical reactions.

Moreover, the reduced sedimentation rate under microgravity conditions affects the distribution of particles in a solution. On Earth, particles in a colloid or suspension will typically settle over time due to gravity, a process known as sedimentation. In the absence of significant gravitational pull, sedimentation is vastly diminished, leading to increased stability in colloidal systems. This stabilization can be particularly advantageous in the synthesis and storage of medicinal products in space, where uniform distribution of nanoparticles in carrier fluids can be maintained for longer durations without the need for frequent agitation.

Microgravity also influences mass transfer processes, which are essential for the rate of many chemical reactions. Diffusion becomes the dominant mechanism for mass transfer when the convective mixing is suppressed. The efficiency of diffusion in microgravity may differ from that under Earth conditions due to changes in density and viscosity of fluids. Researchers have observed that certain reactions proceed at different rates in space as compared to Earth, primarily due to these altered transport phenomena.

The impact of microgravity on phase separation and immiscibility in multiphase systems provides another area of interest. Under normal gravity, immiscible liquids and gases separate rapidly due to differences in density. However, in microgravity, this separation process is much slow, leading to prolonged interactions between phases. This effect can be utilized to enhance the yield of certain chemical reactions where prolonged contact between reactants leads to more complete reaction or different products.

Experimental evidence supporting these theoretical considerations comes from numerous studies conducted on the International Space Station (ISS). For instance, the Zeolite Crystal Growth experiments conducted on the ISS revealed differences in crystal size and morphology when grown in microgravity compared to Earth. Such experiments underline the crucial role of environmental conditions in dictating chemical processes.

Chemical reactions in microgravity not only inform our understanding of fundamental chemistry but also pave the way for innovative applications and technologies in space exploration and beyond. For example, the synthesis of pharmaceuticals in space has been proposed to benefit from microgravity conditions, potentially leading to drugs with fewer impurities and higher efficacy.

Collectively, the effects of microgravity on chemical reactions are significant, affecting both the physical behavior of molecules and the kinetics and thermodynamics of reactions. This necessitates a reexamination of chemical theories and principles through the lens of space-specific factors, an undertaking that could expand the frontier of chemical science into new realms.

1. Calculate the theoretical change in reaction rate of a second-order reaction when conducted under microgravity conditions, assuming other variables remain constant.

2. Design an experiment to test the phase separation of a binary fluid mixture in microgravity. Identify the primary variables that would need to be measured.

3. Discuss the potential impact of altered diffusion rates on the synthesis of a commonly used medication in a microgravity environment.

3.4 Radiation-Induced Chemistry in Space

Radiation-induced chemistry in space represents a crucial area of study, as it significantly affects the chemical makeup and behavior of matter beyond the Earth's atmosphere. In space environments, radiation comes from various sources like cosmic rays, solar wind, and interstellar radiation fields, each contributing uniquely to chemical processes. This exposition delves into the fundamentals of how radiation triggers chemical reactions in space, the types of radiation involved, and their varying impacts on different chemical species.

The primary agents of radiation in space are high-energy particles such as electrons, protons, and heavier nuclei originating from cosmic rays, along with electromagnetic radiation including ultraviolet (UV) and X-rays predominantly emitted by stars including our sun. These radiation forms possess the energy to ionize molecules, leading to the formation of ions and radicals. Such highly reactive species play a pivotal role in initiating and propagating chain reactions, even in the low-density conditions typical of interstellar space.

An important mechanism to consider is the interaction between cosmic rays and interstellar ice, predominantly composed of water, methane, ammonia, and carbon dioxide. Cosmic rays, penetrating these ice mantles, induce ionization and excitation of molecules, thereby facilitating a plethora of reactions that would otherwise be improbable under the frigid and sparse conditions of space. Noteworthy among these is the synthesis of complex organic molecules, including prebiotic compounds. This process is thought to contribute significantly to the chemical richness observed in some regions of space and possibly to the origins of life.

Further, the impact of solar UV radiation on chemical reactions, especially in planetary atmospheres and on the surface of bodies within the solar system, cannot be overstressed. In the upper atmosphere of planets and moons, solar UV radiation drives the photodissociation of molecules, leading to the formation of radical species and initiating complex chemical cycles. For instance, on Mars, the dissociation of carbon dioxide by UV radiation leads to the production of reactive oxygen species, influencing the planet's oxidative environment.

The varying energy spectra of different types of radiation result in a range of effects on chemical dynamics. Higher energy particles can penetrate deeper into dense molecular clouds or planetary surfaces, initiating reactions in regions shielded from lower-energy UV radiation. This stratification of chemical effects contributes to the vertical chemical gradients observed in many celestial bodies.

To analytically approach these phenomena, space missions employ instruments such as spectrometers and mass spectrometers capable of detecting the fingerprints of radiation-induced chemical reactions. Additionally, experimental simulations conducted in lab settings that mimic interstellar and planetary conditions provide valuable insights. These studies involve the use of particle accelerators and UV lamps to simulate cosmic rays and solar radiation, respectively, observing the resultant chemical processes in cryogenically cooled matrices or thin films of substances representative of interstellar ices.

This sophisticated interplay between radiation and chemical matter underscores the adaptability and diversity of chemistry in space. Understanding these mechanisms enhances our comprehension of the cosmos's evolutionary processes and may guide the development of future technologies aimed at exploiting space resources or ensuring the longevity of space missions by managing the deleterious effects of space radiation on materials and systems.

To engage with these concepts, consider the following problems:

1. Calculate the potential yield of complex organic molecules from a given volume of interstellar ice irradiated by cosmic rays, assuming typical cosmic ray flux and ice composition.

2. Discuss how the varying energy spectrum of radiation influences chemical gradients in a densely packed molecular cloud.

3. Analyze the role of radical species produced from UV photodissociation in forming the atmosphere of Mars, proposing experimental setups that could simulate these conditions.

These exercises encourage a deeper engagement with the dynamics of radiation-induced chemistry in extraterrestrial environments, consolidating the theoretical knowledge into practical understanding.

3.5 Chemical Reactions in the Interstellar Medium

The interstellar medium (ISM), the matter that exists in the space between the stars within a galaxy, is predominantly a mixture of

3.5. CHEMICAL REACTIONS IN THE INTERSTELLAR MEDIUM

gases (primarily hydrogen and helium) and dust particles. It offers a unique environment for chemical reactions, significantly different from terrestrial conditions, where extreme temperatures, low densities, and high radiation levels prevail. This section delves into the chemical processes occurring within the ISM, focusing on the formation and transformation of molecular species under these extreme conditions.

Interstellar chemistry is primarily governed by the low density of particles in space. Typical densities in the ISM are about one atom or molecule per cubic centimeter, compared to 10^{19} molecules per cubic centimeter at Earth's surface. This vast difference in density leads to predominantly binary interactions in the ISM, and the probability of three-body interactions, common in terrestrial chemistry, is negligibly small. Therefore, reactions in the ISM often proceed through mechanisms that do not occur or are insignificant under Earth-like conditions.

Formation of molecular hydrogen (H_2), the most abundant molecular species in the ISM, illustrates a key pathway for chemistry in this region. H_2 is primarily formed on the surfaces of dust grains—a process called heterogeneous catalysis. Here, hydrogen atoms adsorb onto the surface of cold dust particles and migrate until they encounter each other, forming H_2 and releasing the molecule back into the gas phase. This mechanism is crucial as H_2 is a major player in the cooling of interstellar clouds and serves as a precursor for more complex molecules.

Radiation also significantly influences chemical reactions in the ISM. The presence of ultraviolet (UV) radiation can lead to photoionization and photodissociation of molecules, forming ions and radicals that are highly reactive. For instance, the radiative environment can ionize carbon monoxide (CO), which then reacts with H_2 to form HCO^+, an important ion in the chemistry of the ISM. These ion-molecule reactions are generally much faster than neutral-neutral reactions at low temperatures, thereby playing a dominant role in the cold regions of the ISM.

In regions with higher density and shielded from intense UV radiation, complex organic molecules (COMs) such as formaldehyde (H_2CO), methanol (CH_3OH), and even more complex species like amino acids can form. The pathways to these molecules usually involve a combination of ion-molecule reactions followed by neutral-neutral reactions. Recent observations and models suggest that ice

CHAPTER 3. CHEMICAL REACTIONS IN SPACE ENVIRONMENTS

mantles on dust grains can serve as reservoirs for many precursors to these COMs. Subsequent warming of these ices, possibly by nearby new stars, can trigger thermal or photo-driven reactions that release these compounds into the gas phase.

An intriguing aspect of interstellar chemistry is the role of isotopic fractionation, which can lead to variations in the isotopic ratios of certain elements (like carbon, nitrogen, and oxygen) compared to their cosmic abundances. Fractionation occurs because isotopes can react slightly differently due to differences in mass. For example, reactions involving ^{13}C might proceed slower than those involving ^{12}C, leading to enrichments or depletions of ^{13}C in different molecular species. This isotopic chemistry provides insights into the physical conditions and chemical history of the interstellar material.

Understanding chemical reactions in the ISM not only reveals the complexity and adaptability of chemistry under extreme conditions but also provides essential information about the processes leading to the formation of stars and planetary systems, and potentially, the origins of prebiotic molecules necessary for life. As observations from telescopes and space missions become increasingly sophisticated, they offer more detailed data to refine our models of interstellar chemistry, enhancing our comprehension of the universe's chemical evolution.

The following problems are aimed at exploring the dynamics of chemical reactions in the ISM:

1. Calculate the reaction rate of HCO^+ formation under typical interstellar conditions, given a CO concentration of 1×10^{-4} relative to H_2 and a typical cosmic ray ionization rate of $1 \times 10^{-17} s^{-1}$.

2. Describe the impact of increasing the dust grain surface area by a factor of ten on the formation rate of H_2. Consider a typical ISM environment with a hydrogen atom density of 100 cm^{-3}.

3. Using isotopic fractionation, explain how ^{13}C enrichments could indicate the evolutionary stage of an interstellar cloud.

3.6 The Role of Temperature and Pressure in Space Reactions

Understanding how temperature and pressure affect chemical reactions is fundamental in terrestrial chemistry; however, in the context of space, these factors present unique challenges and opportunities. In space environments, both temperature and pressure can vary dramatically, from near absolute zero in the voids between stars to the extraordinarily high temperatures and pressures found in stellar interiors or during atmospheric entry to a planet.

Temperature and pressure are critical in determining reaction kinetics, which is the rate at which chemical reactions occur. According to the Arrhenius equation, the rate of a chemical reaction increases exponentially with an increase in temperature. This relationship suggests that chemical reactions in the hotter environments of space, such as near stars or on certain planetary surfaces exposed to solar radiation, proceed much faster than those in colder regions like the outer solar system or shadowed lunar craters.

Conversely, pressure, which is generally defined as the force exerted per unit area, influences the concentrations of reactants in a given volume and thus affects the rate and equilibrium states of reactions according to Le Chatelier's principle. In the near-vacuum conditions of open space or on the surface of small bodies like asteroids, where atmospheric pressure is virtually nonexistent, many reactions that require the collision of particles become less likely or proceed at much slower rates.

The variability of environmental conditions in space significantly affects the chemistry of gases and plasmas. For example, on a planet like Venus, where the surface pressure is about 92 times that of Earth's at sea level and temperatures reach up to 737 K, chemical reactions occur that are not observable under Earth-like conditions. These include the formation of exotic compounds like lead sulfide and bismuth phosphide, predicted from thermodynamic calculations and indicative of the harsh chemical environment.

In the cold and sparse interstellar medium, chemical kinetics play out differently; reactions predominantly occur on the surfaces of dust grains rather than in the gas phase. Here, the dust serves not only as a site for reactions but also as a reservoir for volatile species at low temperatures. These processes are crucial for the formation of complex organic molecules and can be dramatically influenced by minor changes in temperature and pressure, mediated by radiation from nearby stars or energetic events such as supernovae.

Furthermore, experimental simulation of space conditions in laboratories has shown that at low pressures, some chemical processes can produce different products compared to those generated at higher pressures. For instance, the photodissociation of water vapor in microgravity yields a higher proportion of hydroxyl radicals compared to similar reactions conducted under normal Earth-like conditions.

The profound influence of temperature and pressure on space-based chemical reactions offers insights not just into the nature of chemical processes but also into the formation of planetary atmospheres, the evolution of celestial bodies, and the potential habitability of extraterrestrial environments. This knowledge also underscores the importance of accurately assessing these parameters when designing chemical experiments or missions in space, aiming for the synthesis of materials or the extraction of resources.

To further explore these concepts, consider the following application-based questions: 1. Calculate the rate constant of a hypothetical reaction occurring at 300 K and then at 700 K, assuming the activation energy is 50 kJ/mol. How does the rate change with temperature? 2. Given a scenario where a reaction is feasible only above a certain pressure threshold, discuss the feasibility of such a reaction in the vacuum of space versus on the surface of Venus. 3. Predict the products of hydrocarbon reactions in an environment with temperatures exceeding 1000 K and pressures many times that of Earth's atmosphere. What novel chemical structures might be formed under these extreme conditions?

These problems invite students to apply the principles discussed, thereby deepening their understanding of how temperature and pressure influence chemical reactions in the varied and extreme conditions of space.

3.7 Catalysis in Space: Natural and Engineered

Catalysts play a crucial role in numerous chemical processes, both on Earth and in space, facilitating reactions that otherwise would

3.7. CATALYSIS IN SPACE: NATURAL AND ENGINEERED

be inefficient or practically impossible under given conditions. In the unique environment of space, where extreme temperatures, pressures, radiation levels, and the absence of gravity influence chemical reactions, understanding the behavior and application of catalysts becomes even more essential.

Natural catalysis in space environments is evident in processes such as the formation of complex organic molecules in interstellar clouds. These molecules are precursors to life and their synthesis often involves catalytic reactions occurring on the surfaces of cosmic dust or icy comet particles. These natural catalysts, consisting mainly of silicates and carbonaceous compositions, provide surfaces upon which chemical species can adsorb, interact, and transform, under the influence of cosmic radiation and ultraviolet light.

Engineered catalysis, on the other hand, refers to the deliberate design and deployment of catalysts to facilitate specific reactions within extraterrestrial activities, such as habitat construction, life support, and fuel generation in planetary exploration missions. These catalysts must not only be highly efficient and selective but also resistant to deactivation by harsh space conditions such as radiation and temperature extremes.

The unique microgravity conditions of space present both challenges and opportunities for catalysis. On one hand, the reduced gravity affects the distribution of reactants and products near the catalytic surface, potentially leading to increased surface coverage and reaction rates. However, the handling of fluids, critical for delivering reactants through the catalyst bed in many engineered systems, becomes more troublesome. This necessitates the development of new strategies and technologies for effective mass transport in reaction systems.

Thermal management is another crucial consideration. Temperature extremes can significantly influence catalyst activity and stability. Without the natural convection of heat seen on Earth, alternative methods of heat transfer and dissipation need to be engineered to maintain optimal operational temperatures of catalysts.

Radiation-induced catalysis is a unique phenomenon pertinent to space. High-energy radiation can lead to the formation of reactive species that can act as both transient catalysts and reactants. Additionally, radiation can alter the structural properties of catalysts, impacting their functionality. The design of radiation-resistant materials or the incorporation of radioprotective coatings might be necessary to preserve catalytic activity.

One notable case study is the use of titanium dioxide (TiO_2) as a photocatalyst on the International Space Station for air and water purification—an application where its activity is enhanced by the higher levels of UV radiation in space compared to Earth. This demonstrates the potential for certain types of catalysts to exhibit improved performance in the space environment when properly harnessed.

Looking forward, nanotechnology holds promising potential for the advancement of space catalysis. Nanostructured catalysts can offer high surface areas and tunable properties, such as enhanced reactivity and selectivity, tailored specifically for the unique conditions of space environments.

Whether natural or engineered, catalysts designed for space applications must meet stringent criteria of efficiency, stability, and adaptability to extreme conditions. Future research will likely focus on the development of novel catalytic materials that can exploit the unique properties of the space environment, further broadening the realm of possibilities for extraterrestrial chemistry.

3.8 Chemistry on Planetary Surfaces: Mars, Venus, and Beyond

The exploration of planetary surfaces, notably Mars and Venus, provides profound insights into the chemistry that occurs on these worlds, further expanding our understanding of chemical processes beyond Earth's bounds. This exposition will elucidate the unique chemical reactions and conditions on Mars and Venus, whilst considering the prospects of future explorations on similar celestial bodies.

Mars, with its thin carbon dioxide-rich atmosphere and the presence of water ice and permafrost, offers a distinctive ambiance for chemical activities. The oxidation-reduction reactions that are likely occurring have been evidenced by the presence of iron oxides, which give the planet its characteristic red color. The oxidation of iron due to the planet's environmental conditions suggests complex interactions between geological and atmospheric processes. Moreover, the discovery of perchlorates in Martian soil by various rovers points to intriguing oxidative chemistry catalyzed potentially by ultraviolet radiation or electrical discharges such as dust devils and storms. These findings suggest that Mars has an active, if tenuous, photochemically and electrochemically driven set of surface and atmospheric chemical processes.

In contrast, Venus presents a harsher chemical environment with a dense atmosphere primarily composed of carbon dioxide, clouds of sulfuric acid, and extreme surface temperatures exceeding 460 degrees Celsius. The conditions on Venus accelerate chemical dynamics, particularly in the formation of sulfuric acid through the vapor-phase oxidation of sulfur dioxide by photochemically generated radicals under the influence of intense solar ultraviolet radiation. This scenario exemplifies a planetary-scale chemical equilibrium maintained under extreme conditions, driven by volcanic outgassing and continuous recycling through chemical reactions in the atmosphere.

Both planets offer extreme cases for photochemistry and thermochemistry in planetary atmospheres. The interaction of solar radiation with the atmospheric gases triggers a series of photochemical reactions forming complex molecules such as formaldehyde, methane, and even larger organic compounds which have implications for understanding prebiotic chemistry and the origins of life.

One of the intriguing prospects in the study of extraterrestrial surfaces is the potential catalytic activity of minerals. On Mars, for instance, the interaction between silica surfaces and atmospheric gases could facilitate various chemical reactions, paving the way for novel synthetic pathways which are unachievable under Earth's conditions. Similarly, the catalytic roles of mineral oxides and sulfides on Venus must be assessed to understand their contribution to atmospheric chemistry.

The future explorations might focus on transient phenomena like the nightly snows of carbon dioxide on Mars and sulfuric acid on Venus, offering direct insight into dynamic equilibrium and phase changes in non-terrestrial environments. Advanced robotic and spectroscopic technologies are poised to provide more detailed analyses of these reactions and their mechanisms.

For planetary scientists and chemists, these environments are natural laboratories that, while hostile, present unmatched opportunities to study and test our chemical theories and significantly expand the domain of known chemical phenomena. Each mission equipped with more advanced analytical tools brings us closer to a nuanced understanding of extraterrestrial chemistry, thereby laying the groundworks for future colonization or terraforming initiatives.

The chemical investigation of planetary surfaces like Mars and Venus not only augments our understanding of fundamental chemical processes but also enriches our capability to adapt and perhaps manipulate such environments for future human endeavors beyond Earth. This exploration transcends mere academic curiosity—it is a crucial step toward becoming a multiplanetary species.

3.9 Synthesis and Decomposition Reactions in Different Atmospheres

Synthesis and decomposition reactions represent fundamental chemical processes that are pivotal not only on Earth but also across the diverse atmospheres of various celestial bodies. In extraterrestrial environments, these reactions are influenced by unique atmospheric compositions, pressure variations, and temperature extremes, which can dramatically alter reaction pathways and yields.

In the context of planetary atmospheres such as Mars, Venus, and Titan, the synthesis of complex molecules and the decomposition of existing compounds provide critical insight into chemical evolution and potential habitability. Mars, with its thin CO_2-domaninted atmosphere and presence of peroxides and perchlorates in its regolith, exhibits conditions ripe for oxidative decomposition reactions. These reactions can break down organic molecules, posing challenges for microbial life but also offering pathways for generating simpler molecules that can be further transformed into usable energy sources by resilient life forms or via catalytic processes in engineered systems.

Venus, under its dense CO_2 and sulfuric acid cloud cover, showcases another extreme with high-pressure and high-temperature conditions that facilitate unique synthesis reactions not commonly observed on Earth. For example, the formation of sulfuric acid from sulfur dioxide and water vapor is a critical reaction, providing clues to the planetary processes and atmospheric dynamics. Further, the extreme conditions also promote the decomposition of complex organic molecules, which when coupled with high-energy radiation, could lead to the formation of simpler, more stable compounds.

Titan, Saturn's moon, presents yet another intriguing case with its nitrogen-rich atmosphere and hydrocarbon lakes. Here, synthesis reactions involve the formation of tholins — complex organic molecules produced through the interaction of atmospheric methane and nitrogen under the influence of solar ultraviolet radiation. This synthesis is crucial for understanding the chemical basis of prebiotic chemistry in environments outside Earth.

Each of these environments demonstrates unique sets of synthesis and decomposition reactions, influenced largely by the atmospheric conditions prevailing. To study these reactions, in situ experiments, rover-based laboratories, and orbital spectrometers have been deployed. These instruments help identify products and intermediates formed during these reactions and provide valuable data that refine our understanding of extraterrestrial chemistry.

The rate of these reactions is significantly dictated by environmental factors peculiar to each atmosphere. In the low-temperature, high-radiation conditions of Mars, reactions typically proceed at a slower rate compared to Earth, whereas the high temperatures on Venus can accelerate certain chemical reactions. In Titan's cold atmosphere, photochemical synthesis dominates, driven by solar and cosmic radiation.

Theoretical modeling and simulation play crucial roles in interpreting these complex reactions. Computational chemistry tools enable scientists to predict reaction pathways, intermediates, and product distribution under a variety of conditions, thereby extending laboratory findings to conditions that are difficult or impossible to replicate physically. These models require continual refinement to accommodate new data gathered from space missions and laboratory simulations replicating extraterrestrial conditions.

The study of synthesis and decomposition reactions in different atmospheres not only expands our understanding of basic chemical principles but also enhances our knowledge of the potential for life in the universe and informs the design of future space missions aimed at the exploration and possible colonization of new worlds. Chemical insights derived from studying these reactions provide a cornerstone for astrobiology and the future of human technology in space exploration.

The dynamic interplay of synthesis and decomposition reactions in varying atmospheric compositions provides a fascinating lens through which to view chemical and potentially biological processes across the solar system. As we continue to explore the cosmos, the continued study of these reactions will remain a fundamental aspect of unraveling the mysteries of the universe.

3.10 Future Technologies for Controlling Chemical Reactions in Space

In the ambit of extraterrestrial exploration and colonization, controlling chemical reactions within varied and often harsh space environments presents a pivotal challenge. The development of future technologies aimed at managing these reactions underlies not only the sustenance and safety of astronauts but also the feasibility of long-term space habitation and the potential industrialization of space resources.

One promising avenue in this regard is the advancement of microfluidic technology, which enables precise control over the flow and mixing of fluids on a microscopic scale. In microgravity, where traditional methods of fluid mixing are ineffective, microfluidic devices facilitate controlled chemical reactions by manipulating volumes that are orders of magnitude smaller than what is possible with conventional apparatus. This precision significantly minimizes the influence of unwanted variables such as phase separation that occurs in microgravity. Recent experiments aboard the International Space Station (ISS) have shown successful use of microfluidic chips to carry out complex organic syntheses and even biological experiments.

Advancements in automation and robotic assistance form another critical frontier. Robotic systems, equipped with sensors and data-processing capabilities, can adjust reaction conditions in real-time — a task that becomes crucial in dynamic space environments. For example, autonomous systems can modify the pressure and temperature of a reactor vessel on a Mars colony based on atmospheric data sensors collected, thereby ensuring optimal reaction conditions despite the diurnal variations in environmental conditions.

Nanotechnology also offers substantial innovations for controlling chemical reactions in space. Nanocatalysts, due to their high surface area-to-volume ratios, provide enhanced reaction rates and selectivity at lower temperatures and pressures, which are typical in many extraterrestrial settings. These catalysts could be engineered to facilitate specific chemical reactions essential for resource utilization on the moon or Mars, such as extracting oxygen from regolith or synthe-

3.10. FUTURE TECHNOLOGIES FOR CONTROLLING CHEMICAL REACTIONS IN SPACE

sizing vital compounds from the limited available elements.

Quantum computing presents a more nascent, yet potentially transformative, approach by optimizing the planning and execution of chemical reactions. Quantum algorithms are particularly adept at solving complex mathematical problems much faster than classical computers, which includes simulating chemical processes at quantum levels. This capability can be utilized to predict the outcomes of reactions under various conditions or to design molecules that can best utilize the unique conditions of space environments, such as extreme cold or radiation levels.

Energy management technologies also play an essential role. In space, where energy resources are limited, the development of energy-efficient reaction control technologies is critical. Photocatalysis, which uses light to mediate chemical reactions, is particularly appealing. Light-weight solar arrays or possibly even direct cosmic radiation could activate photocatalysts, driving essential reactions without additional energy expenditure.

Finally, the integration of these technologies into closed-loop life support systems in spacecraft or habitats will be imperative. These systems will need to recycle and regenerate resources, managing chemical reactions to ensure the continuous production of water, air, and potentially food. Advanced control systems leveraging machine-learning algorithms could dynamically adjust parameters to optimize resource yield and minimize waste.

These innovative technologies collectively highlight the interdisciplinary approach necessary to tackle the challenges of extraterrestrial chemistry. Each offers a unique set of tools to control and exploit chemical reactions, paving the way for more effective and sustainable human activity beyond Earth.

As space exploration continues to push the boundaries of human reach, our ability to control chemical reactions in space will not only determine the viability of distant explorations but also potentially unlock new paradigms in chemistry that could revolutionize fields beyond space exploration.

Chapter 4

Rocket Propulsion and Fuels

This chapter provides an in-depth analysis of the chemistry behind rocket propulsion systems and the fuels that power them. It discusses the properties and reactions of various chemical propellants used in both liquid and solid rocket engines, detailing how these materials are formulated, stored, and handled to maximize efficiency and safety. The chapter also explores the environmental impacts of rocket propellants and the ongoing efforts to develop greener alternatives that can sustainably support future space exploration missions.

4.1 Introduction to Rocket Propulsion: Principles and History

Rocket propulsion is a fascinating and complex field combining principles from physics, chemistry, and engineering to launch objects into space. The fundamental principle behind rocket propulsion is Newton's third law of motion which states that for every action, there is an equal and opposite reaction. In the context of rocketry, the action is the expulsion of gas out of the engine, and the reaction is the movement of the rocket in the opposite direction.

The history of rocket propulsion begins with the ancient Chinese, who are credited with the development of the first rocket-like devices around the 10th century. These early "fire arrows" were essentially tubes filled with gunpowder that produced a propulsive jet of gases when ignited. By the 13th century, these devices had evolved into rudimentary rocket-propelled weapons used during warfare.

The scientific foundation for modern rocketry was laid in the early 20th century by pioneers such as Konstantin Tsiolkovsky, Robert Goddard, and Hermann Oberth, who each contributed theoretical and practical insights that advanced the field substantially. Tsiolkovsky formulated the rocket equation which quantifies the velocity change achievable by a rocket, depending on the mass of the rocket, the mass of the remaining propellant, and the velocity of exhaust gases expelled by the engine.

Robert Goddard, often regarded as the father of modern rocketry, successfully launched the first liquid-fueled rocket in 1926. His designs and experiments with various types of fuels and engine mechanisms paved the way for the technology used in contemporary rocket engines. Following Goddard, Hermann Oberth and Wernher von Braun further advanced rocket technology during and after WWII, eventually leading to the development of the Saturn V rocket that powered the Apollo missions to the moon.

Chemically powered rocket engines operate on the principle of combustion. Propellants, typically classified into liquid and solid types, undergo rapid chemical reactions in a controlled environment – the combustion chamber. The products of this combustion (primarily gases) expand and are expelled at high speed through a nozzle, creating thrust. The efficiency and strength of thrust are influenced by factors such as the mixture ratio of the propellants, their energetics, and the nozzle design.

Over decades, various chemical propellants have been used based on their performance and storage properties. Early rockets, like those utilized during WWII, primarily used simple solid propellants composed of substances such as nitrocellulose or gunpowder. The transition to liquid propellants provided greater thrust and control over the engine's operations, crucial for larger payloads and longer flights, such as those required in space exploration missions.

Additionally, the shift towards more complex and efficient propellants such as liquid oxygen and kerosene, or liquid hydrogen and liquid oxygen mixtures, marked significant milestones in rocketry. These propellants offered not only higher specific impulses - a measure of thrust per unit weight flow of fuel - but also the versatility needed for missions beyond Earth's orbit.

Today, the principles of rocket propulsion continue to evolve with advancements in technology and materials science. The quest for more efficient, safe, and environmentally friendly rocket systems remains a driving force in the ongoing journey of space exploration. Understanding the basic principles and rich history of rocket propulsion is essential for anyone looking to contribute to, or simply appreciate, this dynamic field of study.

This formatted section offers a comprehensive overview of the principles and historical evolution of rocket propulsion in spac exploration, tailored tot the advanced educational settings.

4.2 Chemical vs. Non-Chemical Propulsion: A Comparative Overview

Rocket propulsion, an essential mechanism for space exploration, can be broadly categorized into chemical and non-chemical systems, each employing distinct principles to achieve thrust. This section delves into the contrasting mechanisms, advantages, and limitations inherent to each type, providing a comprehensive comparative analysis essential for understanding the diverse methodologies used in propelling spacecraft beyond Earth's atmosphere.

Chemical propulsion systems function on the principle of exothermic reactions, wherein chemical propellants undergo controlled combustion in an engine, producing hot gases. These gases expand and escape through a nozzle, generating thrust pursuant to Newton's third law of motion. The chemicals used, typically a fuel and an oxidizer, define the type of chemical rocket engine: liquid, solid, or hybrid. The strength of chemical propulsion lies in its high thrust output, which makes it particularly suitable for the initial stages of space missions when escaping Earth's gravitational pull—an energy-intensive phase.

However, chemical propulsion is not without drawbacks. The requirement for substantial quantities of propellants implies a large mass, which significantly impacts the payload capacity. Moreover, the environmental toll of these propellants, which can include unburnt hydrocarbons, carbon dioxide, and other particulates, raises concerns regarding atmospheric pollution and orbital debris.

In contrast, non-chemical propulsion, including electric and nuclear thermal rockets, presents an alternative by mitigating some of the shortcomings associated with their chemical counterparts. Electric propulsion systems, such as ion thrusters, utilize electric fields to ionize a propellant like xenon, and magnetic fields to accelerate these ions to produce thrust. These systems are characterized by a higher specific impulse—indicating greater efficiency per unit of propellant—than chemical rockets. Electric propulsion's efficiency and lower thrust output make it ideal for tasks requiring fine orbital maneuvers and extended missions, where cumulative thrust over time is viable and desirable.

Nuclear thermal propulsion (NTP), another non-chemical technique, uses a nuclear reactor to heat a propellant like hydrogen, expelling it through a nozzle to generate thrust. NTP systems offer a higher specific impulse compared to traditional chemical rockets and do not depend on solar proximity for energy, making them potentially reliable for deep-space exploration.

The choice between chemical and non-chemical propulsion hinges on mission-specific requirements. Chemical propulsion remains unrivaled for lifting significant payloads from terrestrial surfaces due to its high thrust. For longer missions or those requiring efficient use of minimal propellant mass, non-chemical options become more advantageous. The decision integrates considerations of propulsion efficiency, environmental impact, and mission duration alongside technical and economic feasibility.

Continued advancements in propulsion technology are expected to further refine these systems, with research focused on enhancing the efficiency, reducing environmental impacts, and minimizing the dependency on earth-derived resources. As exploration missions push farther into the solar system, the development and integration of these technologies will play a pivotal role in the success and sustainability of space travel.

This thorough understanding of chemical versus non-chemical propulsion systems not only frames the current landscape of rocket technologies but also sets the stage for predicting future developments in the field of space exploration propulsion.

4.3 Common Chemical Rocket Fuels and Their Characteristics

In rocket propulsion, the choice of fuel is pivotal in determining the performance, efficiency, and viability of both spacecraft and missiles. Chemical rocket fuels, which chemically store energy that is later released in exothermic combustion reactions, are primarily categorized into liquid and solid fuels, each with distinct characteristics and applications.

Liquid rocket fuels generally feature a bipropellant system comprising a fuel and an oxidizer. Among the most commonly used liquid fuels are RP-1, a highly refined form of kerosene, and liquid hydrogen (LH2). RP-1 is favored for its high density and ease of handling at ambient temperatures, making it ideal for first-stage launch engines. It is typically paired with liquid oxygen (LOX) as the oxidizer. This combination was notably utilized in the Saturn V first stage during the Apollo missions, providing a reliable and powerful thrust.

On the other hand, liquid hydrogen offers the highest specific impulse—a measure of how effectively a rocket uses propellant—of any chemical rocket fuel. When combined with liquid oxygen, it produces a highly efficient propulsion system with a specific impulse surpassing that of hydrocarbon-based fuels like RP-1. However, the low density of liquid hydrogen presents challenges in fuel tank design, requiring more voluminous tanks to hold an equivalent amount of energy, which can be a disadvantage in terms of structural weight and balance.

Solid rocket fuels are composed of a binder, oxidizer, and a fuel component mixed into a homogeneous propellant mass that is cured into a solid grain. An example of such a fuel is the ammonium perchlorate composite propellant (APCP), which consists of a mixture of ammonium perchlorate as the oxidizer, powdered aluminum as the fuel, and a polymer like HTPB (Hydroxyl-terminated polybutadiene) as the binder. These fuels are celebrated for their simplicity and reliability, offering immediate thrust which is highly advantageous in applications requiring quick acceleration, such as missile launches or escape systems.

Both liquid and solid rocket fuels are characterized by their distinct storage and handling requirements. Liquid fuels require complex, heavy, and insulated storage tanks as well as mechanical pumps for fuel delivery to the combustion chamber. Solid fuels, while simpler in architecture, involve intricate processes in the casting of the propellant grains to ensure uniformity and stability. These grains must also be vigilantly monitored for defects that could lead to catastrophic failure.

Another important aspect of rocket fuels is their stoichiometry—the molar ratio of oxidizer to fuel—which must be precisely controlled to optimize the energy output and efficiency of the rocket engine. Incomplete combustion can lead to diminished thrust and increased residuals, undermining both performance and reliability.

The choice of rocket fuel influences not just the engineering design of the rocket and its launch infrastructure, but also operational strategies. Factors like Earth's gravitational pull, desired payload weight, destination (orbital, interplanetary or beyond), and mission duration heavily dictate fuel selection. For instance, missions requiring high velocities and prolonged propulsion, such as trips to Mars, might opt for high-energy-density propellants like LH2/LOX despite the associated technical challenges.

Chemical rocket fuels play an undeniably crucial role in the field of rocket propulsion, each type bringing a unique set of properties that cater to different facets of aerospace engineering. The continuous evolution and rigorous optimization of these fuels remain fundamental to advancing space exploration ambitions.

4.4 Chemistry of Combustion in Rocket Engines

Rocket propulsion relies fundamentally on the exothermic reactions that occur during the combustion of propellants. At its core, the chemistry of combustion in rocket engines involves complex interactions between oxidizers and fuels, leading to the production of high-energy exhaust gases that propel the rocket forward. This section delves into the stoichiometry of these reactions, the role of catalysts, and the resulting thermochemical phenomena characteristic of rocket engines.

Combustion in rocket engines is primarily a rapid oxidation process of the fuel by the oxidizer. Liquid rocket engines commonly use bipropellant systems, whereas solid rocket engines use a pre-mixed

4.4. CHEMISTRY OF COMBUSTION IN ROCKET ENGINES

composition of oxidizer and fuel. The choice of fuel and oxidizer and their correct proportioning dictates the efficiency, thrust, and stability of the rocket engine.

Hydrogen and oxygen are a common bipropellant duo in liquid rocket engines, exemplified by their use in the Space Shuttle main engines. The stoichiometry of the hydrogen-oxygen reaction in a liquid rocket engine is typically balanced as follows:

$$2H_2 + O_2 \rightarrow 2H_2O + \text{Energy}$$

This reaction produces water vapor and releases a substantial amount of heat, which translates into high specific impulse—a measure of propulsion efficiency. The temperature of the combustion chamber can approach 3500 K, necessitating the use of advanced materials and cooling methods to withstand and manage the extreme thermal environment.

In solid rocket engines, the propellant is a mixture of a powdered oxidizer (like ammonium perchlorate) and a polymer fuel (such as HTPB—Hydroxyl-terminated polybutadiene), often with added metallic fuels like aluminum to increase the energy density. The combustion reaction in a typical solid rocket might be represented as follows:

$$6NH_4ClO_4 + 5Al + 6HTPB \rightarrow 4Al_2O_3 + 12CO_2 + 12H_2O + 6HCl + \text{Energy}$$

This reaction is not only crucial for generating thrust but also must be carefully controlled to regulate burn rate and pressure within the combustion chamber.

To accomplish precise control over these vigorous reactions, rocket engines often utilize catalysts. For instance, in hypergolic rocket fuels—which ignite spontaneously upon contact between fuel and oxidizer—catalysts play a vital role in stabilizing the reaction rate and controlling energy release. Commonly used hypergolic propellants include the combinations of hydrazine (N_2H_4) with nitrogen tetroxide (N_2O_4).

The efficiency of combustion processes in rocket engines is also critically dependent on the specific conditions under which they operate. Parameters such as pressure, temperature, and mixture ratio (the ratio of oxidizer to fuel) are optimized based on the principles of chemical kinetics and thermodynamics. Notably, the performance of the rocket engine is evaluated through the exhaust velocity, which is derived from the kinetic energy of the reaction products according to the ideal rocket equation:

$$v_e = \sqrt{\frac{2kRT}{M}}$$

where v_e is the exhaust velocity, R is the universal gas constant, T is the absolute temperature of the exhaust gases, M is the molar mass of the exhaust gases, and k represents the specific heat ratio.

Understanding the subtleties of rocket engine combustion not only involves evaluation of the chemical reactions themselves but also necessitates a thorough examination of the resulting environmental impacts, such as emissions of CO_2, HCl, and particulate matter. Thus, rocket scientists and engineers continuously strive to optimize the chemistry involved, seeking fuel and oxidizer combinations that maximize thrust while minimizing adverse environmental and operational impacts.

The chemistry of combustion in rocket engines is a vibrant field of study that integrates principles from several domains of science and engineering to ensure that rocket propulsion is both efficient and effective. As the field advances, ongoing research focuses on uncovering new propellant formulations and combustion techniques that could further enhance the performance and sustainability of rocket propulsion.

4.5 Liquid Fuels: Composition, Storage, and Handling

Liquid rocket fuels, or liquid propellants, are a pivotal component in modern rocketry, offering distinct advantages over their solid and hybrid counterparts in terms of efficiency, controllability, and performance. These fuels typically consist of a fuel and an oxidizer, which, when combined, undergo a chemical reaction to produce thrust. This section delves into the chemical composition of various liquid fuels, and outlines the intricacies of their storage and handling to ensure safety and optimal performance in rocket propulsion.

Rocket engines that utilize liquid fuels function on the basis of controlled chemical reactions between the chosen fuel and oxidizer. Two primary types of liquid propellant combinations are commonly used: hypergolic propellants and cryogenic propellants. Hypergolic propellants, such as hydrazine (N_2H_4) paired with nitrogen tetroxide

4.5. LIQUID FUELS: COMPOSITION, STORAGE, AND HANDLING

(N_2O_4), ignite spontaneously on contact with each other at room temperature, eliminating the need for an ignition system. This feature makes hypergolic engines highly reliable and simple in mechanical design, suitable for spacecraft maneuvers including adjustments in orbit and lunar landings. However, these chemicals are highly toxic and pose significant handling risks.

On the other hand, cryogenic propellants like liquid hydrogen (LH_2) and liquid oxygen (LOX) must be stored at extremely low temperatures to maintain their liquid state. LH_2 is maintained at approximately 20 K ($-253°C$) and LOX at about 90 K ($-183°C$). These substances offer the highest efficiency in terms of specific impulse—a measure of how effectively a rocket uses the mass of its propellants. However, the cryogenic nature of these fuels requires sophisticated insulation technologies to prevent heat ingress and fuel evaporation. The storage tanks and fuel lines must also be manufactured with materials that can withstand the low temperatures without becoming brittle.

The storage systems for these liquid propellants need to be meticulously designed to prevent accidental release or degradation of propellants. For ground storage, large insulated tanks with stringent safety mechanisms are utilized, while in-space designs must consider the added complexities of microgravity. In addition to physical storage, the transfer of these fuels into the rocket's tanks is a critical process that necessitates precision. Automated systems are often employed to handle the transfer, measuring, and mixing of propellants to exact specifications determined by mission requirements.

Handling procedures for liquid rocket fuels are governed by strict safety protocols due to the reactive and often toxic nature of these substances. Protective equipment is mandatory, and all personnel must undergo extensive training focusing on dealing with spills, leaks, and fire hazards. Emergency response tactics are planned and rehearsed to deal with potential accidents, ensuring swift resolution while minimizing risk to personnel and infrastructure.

Environmental control systems are also essential when dealing with volatile or cryogenically stored fuels to monitor for any leaks of toxic or ozone-depleting substances which could impact local ecosystems. These systems are integrated with automatic shutdown and containment procedures that activate upon detection of a leak, further securing the operation environment.

The selection, storage, and handling of liquid fuels in rocketry are defined by a balance of efficiency, safety, and environmental considerations. Each type of liquid fuel brings its own set of challenges and benefits, influencing their use in various applications in space exploration. As we continue to push the boundaries of space travel, the development of more advanced liquid propulsion systems, coupled with improvements in safety and environmental practices, will play a crucial role in enabling more ambitious missions beyond our current horizons.

4.6 Solid Rocket Fuels: Advantages and Composition

Solid rocket fuels, commonly known as solid rocket propellants, provide a distinct set of advantages over their liquid and hybrid counterparts, which makes them an invaluable component in various rocket propulsion systems. Fundamental to the operational simplicity of solid rocket fuels is their premixed composition of oxidizer and fuel, which exists in a solid state. This attribute allows for a more straightforward and robust design of rocket engines, as the need for complex fuel delivery systems is eliminated.

The composition of solid rocket propellants typically involves a binder, which doubles as a fuel, an oxidizer, and various additives that enhance performance or provide specific properties such as stability and smoke reduction. The binder, often a synthetic rubber such as hydroxyl-terminated polybutadiene (HTPB), not only provides structural integrity to the fuel but also contributes to the combustion process. The primary oxidizer used in these fuels is often ammonium perchlorate (AP), which decomposes at high temperatures releasing oxygen to sustain the combustion of the binder.

The strategic selection of additives plays a pivotal role in the performance of solid rocket fuels. For example, aluminum is frequently used as a fuel additive due to its ability to increase thrust and improve the specific impulse of the rocket. Burning at high temperatures, aluminum adds to the total heat output, enabling more efficient combustion processes. Furthermore, stabilizers and plasticizers may be included to enhance the mechanical properties of the propellant and its overall stability during storage and handling.

One of the most compelling advantages of solid rocket propellants lies in their ability to be cast into various shapes and sizes, tailored

4.6. SOLID ROCKET FUELS: ADVANTAGES AND COMPOSITION

to the specific thrust requirements of the mission. This molding capability allows for the design of grains, which are geometries of the solid propellant within the rocket motor, to control the burning surface area over time, thus regulating the thrust profile of the launch vehicle. This feature is critical in missions requiring precise thrust control without the complexity of throttleable engines.

Moreover, solid rocket fuels are recognized for their safety and stability in storage and handling compared to liquid fuels. Without the need for cryogenic storage or complex pressurized containers, solid propellants can be maintained and transported with reduced risk. This factor is especially significant in military applications, where the ruggedness and readiness of the propulsion system are paramount.

From an environmental standpoint, challenges remain with the use of traditional solid rocket propellants, particularly concerning the release of hydrochloric acid from the decomposition of ammonium perchlorate. This issue underscores the importance of ongoing research into alternative oxidizers and binders that can minimize the ecological footprint of solid rocket fuels. Recent advancements have seen the exploration of more eco-friendly materials such as ammonium dinitramide (ADN), which promises a lower production of harmful byproducts.

Solid rocket fuels' inherent stability, ease of handling, and capacity to be engineered to meet specific mission requirements make them an enduring choice for a wide range of aerospace applications. Despite environmental challenges, innovation in propellant formulation continues to enhance the viability of solid rocket fuels in the realm of modern rocketry. Further investigation into green propellant alternatives will remain essential in aligning the technology with the broader goals of sustainability in space exploration.

Problems

- Calculate the theoretical specific impulse of a solid rocket fuel composed of 68% ammonium perchlorate, 18% aluminum, and 14% HTPB. Assume complete combustion and standard atmospheric conditions.

- Discuss the impact on performance characteristics like burn rate and specific impulse when substituting hydroxyl-terminated polybutadiene (HTPB) with polyethylene glycol in solid rocket propellant formulations.

- Evaluate the environmental impact of using ammonium perchlorate as an oxidizer in solid rocket propellants and propose potential alternative materials with lower ecological footprints.

- Design a grain configuration for a mission requiring a solid rocket motor to operate in a three-stage burn: initial peak thrust, followed by a reduced steady thrust, and concluding with a final peak. Illustrate your design and describe the role of the grain geometry in achieving the desired thrust profile.

4.7 Hybrid Rocket Fuels: Combining the Best of Both Worlds

Hybrid rocket propulsion systems represent a sophisticated technology that amalgamates the advantageous features of both solid and liquid rocket fuels, aiming to optimize performance, safety, and environmental sustainability. Hybrid rockets utilize a solid fuel in combination with a liquid or gaseous oxidizer. This distinctive configuration facilitates a controlled combustion process, which is inherently safer and more stable than those found in purely solid or liquid propulsion systems.

The architecture of a hybrid rocket engine involves a combustion chamber containing the solid fuel, typically a polymer such as hydroxyl-terminated polybutadiene (HTPB), paraffin wax, or rubber-like substances. The oxidizer, often liquid oxygen, nitrous oxide, or hydrogen peroxide, is stored separately and only introduced into the combustion chamber during operation. This separation of fuel and oxidizer until the point of combustion fundamentally enhances

4.7. HYBRID ROCKET FUELS: COMBINING THE BEST OF BOTH WORLDS

the safety characteristics of hybrid rockets by reducing the risk of accidental ignition.

Chemically, the reaction within a hybrid engine is governed by surface area contact between the oxidizer and the fuel. As the oxidizer is injected over the solid fuel surface, it decomposes or reacts to produce gases that heat and consequently decompose the solid fuel. This decomposition produces further gas, which expands and is expelled to produce thrust. The control of the oxidizer flow rate directly affects the thrust output and burn rate, enabling throttleable propulsion which is highly advantageous for precise maneuvers in space missions.

The advantages of hybrid rocket engines extend beyond safety and control. From an environmental standpoint, hybrids typically emit fewer particulates and have a lower carbon footprint compared to traditional solid propellants. The choice of fuel and oxidizer can significantly influence the environmental impact. For instance, using liquid hydrogen and oxygen leaves only water as a combustion byproduct, marking it as an exceptionally clean propulsion method.

However, the practical application of hybrid propulsion is not devoid of challenges. The design and manufacturing of effective hybrid rocket engines require meticulously engineered components to manage the disparate physical states and interaction dynamics of the solid fuel and liquid oxidizer. Scale-up issues such as inconsistent fuel regression rates, oxidizer leakage due to differential thermal expansion, and ignition reliability are significant engineering hurdles.

In-depth studies and experimental launches have illustrated these challenges. For example, the SpaceShipTwo of Virgin Galactic employs a hybrid rocket engine using HTPB and nitrous oxide. The scalability of this technology was tested during development phases, highlighting issues like the variability in fuel grain combustion, which needed substantial engineering insights to rectify.

Continued research focuses on optimizing fuel formulations to enhance regression rates without compromising stability. Advances in materials science offer promising avenues, such as incorporating nano-additives into the solid fuel to increase the surface area and thus the burn efficiency. Another area of active research is the development of advanced injectors for the oxidizer to improve mixing and combustion stability at various scales of operation.

Hybrid rocket fuels embody a significant stride towards safer, more efficient, and environmentally friendly space exploration. Their capacity to combine the benefits of solid and liquid propellants opens up new possibilities for rocket design and mission planning. As technological and material advancements continue, the efficacy and applications of hybrid rockets are set to expand, underpinning future propulsion systems in the aerospace industry.

Problems

Problem 1: Design a small-scale hybrid rocket engine for a university project. Outline the choice of solid fuels and oxidizers, considering the trade-offs between combustion efficiency, safety, and environmental impact.

Problem 2: Calculate the theoretical specific impulse achievable with a hybrid rocket using paraffin wax as fuel and liquid oxygen as the oxidizer. Assume complete combustion.

Problem 3: Discuss the implications of varying oxidizer flow rate in a hybrid rocket engine in terms of thrust control and stability during flight.

Problem 4: Evaluate the environmental impacts of using nitrous oxide as an oxidizer in terms of greenhouse gas emissions and ozone depletion potential. Compare this with the impacts of using liquid oxygen.

4.8 Environmental Impact of Rocket Fuels

The utilization of rocket fuels in the realm of aerospace propulsions has played an indispensable role in enabling humanity to access, explore, and utilize space. However, it has also introduced numerous environmental impacts that necessitate comprehensive study and ameliorative approaches. Among these impacts, atmospheric pollution, ozone layer depletion, and contribution to climate change are some of the most critical.

Rocket launches are inevitable sources of atmospheric pollutants. During ignition, rocket engines release a myriad of combustion products, primarily carbon dioxide (CO_2), water vapor (H_2O), carbon monoxide (CO), hydrochloric acid (HCl), unburnt hydrocarbons, and various nitrogen oxides (NO_x). For instance, solid rock-

4.8. ENVIRONMENTAL IMPACT OF ROCKET FUELS

ets burning a composite of ammonium perchlorate and powdered aluminium emit copious amounts of HCl, which, once released into the atmosphere, can lead to the formation of acidic particulates detrimental to both terrestrial ecosystems and human health. Further, the voluminous release of CO_2 and water vapor by rockets contributes significantly to the global greenhouse effect, an environmental syndrome associated with rising global temperatures.

Critical to understanding the environmental repercussions of rocket propulsion is the study of the upper atmospheric effects, most notably on the ozone layer. The ozone layer is vital in protecting Earth from ultraviolet radiation. Rockets directly deposit reactive chemicals like chlorine and bromine into the upper atmosphere, which can potentially disrupt the ozone layer. Chlorine atoms from HCl (from solid rocket motors) and bromine from certain brominated hydrocarbons used in a few rocket propellants, actively participate in ozone depletion processes. Catalytic reactions involving these halogens break down ozone (O_3), converting it back to molecular oxygen (O_2), diminishing the ozone layer's protective capacity against harmful solar radiation.

The geographical area of impact also plays a crucial role in the environmental footprint of rocket launches. The sites chosen for rocket launches are often located near coastlines to minimize human risk in the event of a launch failure. However, these regions can be ecologically sensitive, hosting a wide variety of endemic species and unique ecosystems such as mangroves and coral reefs. For example, the long-term effects of frequent launches from facilities like Florida's Kennedy Space Center on the adjacent coastal ecosystems are under intensive study to delineate the precise ecological impacts, which include not only chemical pollution but also acoustic and thermal disturbances associated with launches.

Given the escalating frequency of orbital and suborbital flights, driven by both national space agendas and private aerospace ventures, the scale of these environmental effects is increasingly significant. It has catalyzed the development of stringent regulations regarding rocket emissions and has spearheaded research into alternative, less harmful propellant chemistries. Innovations in green propulsion technologies, such as the use of liquid methane and hydrogen with oxygen—which primarily yield water as a combustion product—are seen as promising pathways to reduce these environmental detriments.

Moreover, efforts are underway to enhance the environmental friendliness of rocket operations, not merely through propellant reformulation but also via improvements in launch architecture and post-launch cleanup strategies. For instance, recapture and recycling of propellant remnants, integration of environmental impact assessments in the preliminary design of new rocket systems, and robust regulation compliance monitoring are all vital components of a sustainable space exploration agenda.

The imperative to reduce the environmental footprints of rocket fuels while maintaining or enhancing their propulsive capabilities is a key challenge for chemical engineers and environmental scientists. This dual objective requires an interdisciplinary approach, combining insights from chemistry, materials science, and environmental science to innovate solutions that can sustain the burgeoning demand for space exploration while safeguarding our planet's environmental health. As this field progresses, continuous monitoring, adaptive management of space launch activities, and comprehensive environmental impact research will be essential in ensuring that the pursuit of exploring beyond our atmosphere does not come at the expense of its integrity.

4.9 Innovations in Green Propulsion Technologies

The pursuit of environmentally friendly propulsion technologies has transformed the landscape of chemical rocket propellants. As the space industry accelerates toward more sustainable practices, significant advancements have been made in developing green propulsion technologies. These innovations not only promise to reduce the environmental footprint of rocket launches but also enhance the efficiency and cost-effectiveness of space missions.

A primary focus has been the development of less toxic and more eco-friendly rocket fuels. Among these, the replacement of hydrazine, a commonly used but highly toxic propellant, with greener alternatives stands out. Hydroxylammonium nitrate (HAN) is an emerging option; when formulated as an ionic liquid, it functions as a high-performance monopropellant that offers lower toxicity and comparable performance without the carcinogenic risks associated with hydrazine. Additionally, the combustion products of HAN-based fuels—chiefly water vapor and nitrogen—exhibit minimal environmental impact, which significantly eases the burden on terrestrial and atmospheric ecosystems.

Further advancements have involved the utilization of bio-derived fuels, which are obtained from renewable resources. One promising class of materials involves bio-derived alcohols, such as ethanol and butanol, utilized in formulations that leverage their biodegradable nature. When these biofuels combust, they produce significantly lower amounts of soot and greenhouse gases compared to traditional petroleum-based rocket fuels. The adaptability of bio-alcohols to existing liquid rocket technology frameworks also promotes their integration into current propulsion systems with minimal retrofitting required.

Another innovation in the sphere of green propulsion is the use of solid rocket fuels incorporating environmentally benign oxidizers. Ammonium dinitramide (ADN) is one such oxidizer that has gained attention. In solid fuel formulations, ADN offers a lower environmental impact due to its clean burning properties and stability, reducing the production of hazardous waste products.

The technology of cryogenic and supercritical fluids as propellants also presents notable green alternatives. Liquid oxygen, a well-established cryogenic liquid, is being explored further to optimize its performance in combination with renewable fuels; meanwhile, supercritical carbon dioxide has been researched as a working fluid in propulsion systems. These methods not only diminish toxic emissions but are also characterized by their reusability and reduced ecological disturbance.

Progress in propulsion is not limited to the chemicals themselves but extends to the design and operation of rocket engines. Electrochemical engines, for instance, use electricity to generate propulsion from certain chemical reactions that are inherently cleaner than traditional combustion processes. This approach not only alleviates the release of noxious emissions but potentially allows for the recapture and reuse of some byproducts.

Advancements in green propulsion are anticipated to accelerate further as regulatory demands tighten and as the push for sustainable space exploration grows stronger. The integration of these technologies not only serves to fulfill environmental commitments but also enhances the propulsion efficiencies needed for longer and more demanding missions, thereby supporting a sustainable trajectory for future space exploration.

In summary, these innovations in green propulsion technologies paint a promising picture of a future where space exploration aligns seamlessly with environmental stewardship. The ongoing research and development in this field are crucial in minimizing the ecological impact of humanity's quest to explore beyond our planet.

4.10 Future Trends in Rocket Fuels and Propulsion Systems

As the landscape of space exploration evolves, the development of advanced rocket fuels and propulsion systems is critical to enable more efficient, sustainable, and far-reaching missions. This section explores the anticipated trends and innovations that promise to redefine rocket propulsion in the coming decades.

One of the primary areas of focus in future rocket propulsion is the enhancement of fuel efficiency. Advanced formulations of both liquid and solid fuels aim to deliver higher specific impulse—a measure of how effectively a rocket uses fuel. For example, the use of boron or lithium as additives in traditional propellants could potentially increase the energy density, thus providing a higher specific impulse compared to conventional chemical propellants alone.

In addition to improving traditional fuels, significant research is being invested into alternative propellants that are less harmful to the environment. One promising alternative is the development of bio-derived rocket fuels, which are produced from renewable resources. These biofuels aim to reduce the carbon footprint of rocket launches and potentially offer better performance characteristics due to their cleaner burning properties. Researchers are experimenting with various bio-derived compounds, such as synthesized paraffinic kerosene, which can be produced from algae and other bio-oils.

Another innovative approach in the propulsion systems of the future is the application of nuclear thermal propulsion (NTP). This technology leverages nuclear reactions to heat a propellant, hence expelling it at higher velocities than traditional chemical rockets. NTP could significantly reduce travel time to distant celestial bodies like Mars, enhancing the feasibility of manned missions beyond the Moon. Although challenges related to safety and radioactive emissions remain,

4.10. FUTURE TRENDS IN ROCKET FUELS AND PROPULSION SYSTEMS

ongoing advancements in reactor design and control are expected to make NTP a viable option for deep space exploration.

Electrical propulsion systems, such as ion thrusters and Hall effect thrusters, are receiving increased attention for their applications in longer-duration missions. These systems utilize electric fields to accelerate ions, providing a much higher efficiency compared to conventional chemical rockets. Although currently limited by low thrust output, ongoing research aims to enhance power densities and thrust capabilities, which could make electric propulsion systems ideal for tasks like orbital adjustments and interplanetary travel when coupled with chemical launch systems.

On the forefront of experimental propulsion methods, antimatter and fusion-based systems represent the most radical departures from conventional technologies. Antimatter propulsion, theoretically, offers the highest energy density and efficiency possible. However, challenges related to antimatter production, storage, and controlled annihilation must be addressed before such systems can be practical. Similarly, fusion propulsion promises significant improvements in efficiency over current technologies by mimicking the energy production processes of stars. Researchers are exploring various fusion techniques, including magnetic confinement and inertial confinement, to harness this powerful energy source safely and efficiently.

The evolution of propulsion technologies continues to be supported by advancements in materials science. New composite materials and manufacturing technologies, such as 3D printing, are being developed to create components that are lighter, stronger, and capable of withstanding the extreme conditions of space and the high temperatures of advanced propulsion systems.

As we look to the future, the intersection of chemical engineering, materials science, nuclear physics, and environmental science will play a pivotal role in shaping the next generation of rocket fuels and propulsion technologies. These advancements will not only enhance the capabilities of space missions but also contribute to more sustainable practices in rocket propulsion, aligning with broader environmental objectives and the long-term goals of human space exploration.

Chapter 5

The Role of Water and Ice in Space Chemistry

This chapter examines the pivotal role that water and ice play in the chemical processes of space environments. It discusses the distribution of water and ice across the solar system, including their presence on planets, moons, and comets, and how these substances participate in various chemical reactions that may support life. The text also covers the methodologies for detecting and analyzing water and ice in celestial bodies, the use of water in life support systems within spacecraft, and the potential of water as a resource for future space missions.

5.1 Importance of Water and Ice in Space Exploration

Water and ice play indispensable roles in space exploration, serving as a cornerstone for supporting life, facilitating the generation of fuel, and contributing to the sustenance of space habitats. These substances are not only imperative for current space missions but are also crucial for the future colonization of other planets.

Water, with its unique chemical and physical properties, acts as one of the most vital resources in space. Its polarity and ability to form hydrogen bonds make it an excellent solvent, which is essential for any biochemical process. The presence of liquid water is one of the primary criteria in the search for extraterrestrial life. The biological importance of water arises from its involvement in the synthesis and function of the macromolecules required for life. Thus, the detection of water in extraterrestrial environments suggests potential habitats that might support life forms similar to those on Earth.

In the context of manned space missions, water is crucial not only for drinking but also for providing oxygen and hydrogen through the process of electrolysis. The hydrogen produced can be used as fuel, while the oxygen is indispensable for respiration. Furthermore, water serves as a heat sink in spacecraft systems, helping manage the internal temperature and maintain the functionality of critical components.

Ice in the extraterrestrial environment, particularly on moons and distant planets, is equally significant. Ice caps and glaciers have been discovered on Mars, and water ice is abundant on several moons of the outer planets, including Europa, Enceladus, and Titan. The presence of ice is a critical indicator of water reservoirs that could be tapped into future missions. Extracting this ice could supply long-term human expeditions or autonomous robotic operations with necessary resources, reducing the need to transport large quantities of water from Earth—a process that is both cost-prohibitive and technically challenging.

Moreover, ice plays a vital role in the geology and climatology of celestial bodies. It influences surface albedo, atmospheric composition, and even geological activity, as seen in cryovolcanism on some icy moons. Understanding these processes is essential not only for comprehending the celestial body itself but also for planning human activities and establishing habitats in these environments.

Additionally, water is considered for its potential to shield against space radiation. Water's high hydrogen content makes it an effective barrier against galactic cosmic rays and solar particles, which pose serious health risks to astronauts on long-duration space missions. Implementing water-based radiation shielding could be more effective and less costly than other materials currently used.

Economic and logistical considerations also underscore the importance of water and ice in space exploration. Utilizing in-situ resources (ISRU) for producing fuel and life-support requirements significantly reduces the payload mass, launch costs, and dependence on Earth-based resources. The concept of using lunar or Martian water for creating fuel and supporting life is a cornerstone in the strategy for sustainable, long-term human presence beyond Earth.

The multi-faceted roles of water and ice extend across scientific research, practical applications in life support systems, and the strategic management of future space missions. Understanding and harnessing these resources efficiently is pivotal to advancing human exploration and potentially inhabiting other celestial bodies. Thus, water and ice are not merely subjects of scientific inquiry but fundamental resources that might shape the future of human space exploration.

5.2 Physical and Chemical Properties of Water and Ice in Space

The exploration of physical and chemical properties of water and ice in extraterrestrial environments is critical for understanding their roles and behaviors in various celestial settings. Water (H_2O), a ubiquitous substance across the cosmos, exhibits unique properties that can radically differ under space conditions compared to those on Earth due to variations in pressure, temperature, and radiation exposure.

Phase Transitions and Behavior in Microgravity

In the vacuum of space and on celestial bodies with thin atmospheres, water mainly exists in the form of ice or vapor, as the low pressures generally prevent the existence of liquid water. The phase diagram of water alters accordingly due to lower pressure environments. For example, on Mars, due to reduced atmospheric pressure, water ice sublimates directly into water vapor without transitioning through a liquid phase. This phase behavior is crucial for understanding atmospheric processes and surface-interaction phenomena on Mars and other similar bodies.

Microgravity, a condition frequently encountered in space travel and on small celestial bodies, significantly affects the behavior of water and ice. In microgravity, surface tension dominates over other forces, causing water to form cohesive spherical droplets. These unique behaviors affect not only basic physical processes but also influence the design of life support and thermal control systems in spacecraft.

Crystal Structure and Variability

In space conditions, especially on the cold surfaces of moons and distant planets, water ice can exist in multiple crystalline forms. The most common on Earth, hexagonal ice (Ih), might be accompanied or even replaced by cubic ice (Ic) or amorphous ice under varying temperatures and pressures found in extraterrestrial environments. This variability in crystal structure can impact the reflectivity and emissivity of icy surfaces, influencing thermal balance on planetary surfaces and the interpretation of spectroscopic data from remote sensing missions.

Radiation Effects on Water and Ice

Radiation significantly influences the chemical properties of water and ice in space environments. High-energy particles, such as those from solar wind and cosmic rays, can induce radiolysis, breaking water molecules into hydrogen and hydroxyl radicals. These products can recombine or react with other substances, influencing chemical pathways and potentially contributing to the creation of complex organic molecules. Biological implications are profound as radiation-altered ice may offer both shelter from radiation and a chemically reactive environment potentially suitable for supporting or generating life.

Amorphous Ice and Clathrate Hydrates

Amorphous ice, a non-crystalline form of ice, is particularly relevant in cold, outer regions of the solar system and in dense molecular clouds of space where temperatures are exceptionally low. Due to its disordered structure, amorphous ice may trap gases such as carbon dioxide, methane, or ammonia, forming clathrate hydrates. These structures are essential for considering the potential storage and transport of gases across the solar system and may serve as a resource or challenge in future space exploration and colonization.

Understanding the physical and chemical properties of water and ice under varied space conditions not only informs theoretical models and observational techniques but also serves practical applications in space exploration missions. From the management of water resources on spacecraft and future colonies to the utilization of local water sources on moons or Mars, addressing the unique challenges posed by space environments is crucial for successful long-duration space exploration. The fundamental chemistry and physics of water and ice guide essential decisions from spacecraft design to mission planning and execution, underscoring their significance in the broader context of astrochemistry and the pursuit of life beyond Earth.

5.3 Water and Ice in the Solar System: Distribution and Sources

The discovery and study of water and ice in the solar system is a critical aspect of modern astronomy and astrochemistry, affiliating directly with the feasibility of life, geology, and potential human colonization in space. Water (H_2O) exists in various forms across the solar system, from vapor and liquid brines to massive crystalline ice structures and frost. The sources of water and ice are as diverse as their distribution, creating significant implications for our understanding of planetary formation, dynamics, and sustainability of possible life forms outside of Earth.

One of the prime celestial bodies containing water is Earth, where water is abundant in all three phases: solid, liquid, and gas. Moving beyond Earth, the presence of water ice has been extensively recorded on the Moon. Recent missions, like the Lunar Reconnaissance Orbiter and Chandrayaan-1, have discovered water ice in permanently shadowed craters near the lunar poles. These craters, shielded from the sun's heat, serve as natural cold-traps gathering and preserving ice - a potential resource for lunar bases.

Mars, with its historic abundance of water, presents evidence of hydrated minerals and ancient river valleys. Furthermore, the Mars Reconnaissance Orbiter has provided substantial proof of subsurface ice, and the polar caps comprised primarily of water ice, coated by a thin layer of dry ice (frozen carbon dioxide) during the winter.

In the outer solar system, Jupiter's moons show a fascinating diversity of icy environments. Europa is one of the most promising locations for extraterrestrial life because of its subsurface ocean, which lies beneath a thick ice crust. This indication of vast liquified water combined with the plume activity observed by the Hubble Space Telescope suggests active exchange between the icy surface and the inner ocean, enriching the potential for biological activity. Ganymede, the largest moon in the solar system, has an icy crust and likely a subterranean ocean. Data from the Galileo spacecraft have indicated that its icy surface is dynamically young and has been continually resurfacing, a process that might involve internal oceans and possible water reservoirs.

Saturn's moon Enceladus is another key player in solar system hydrology. It has been observed spouting huge plumes of water vapor and ice from its southern hemisphere via "tiger stripes", which are tectonic fractures. The Cassini spacecraft has sampled these plumes, suggesting a subsurface ocean rich in salts and organic compounds, making Enceladus a prime target for astrobiological studies.

Moving further into the Kuiper Belt, Pluto, as observed by the New Horizons mission, exhibits a vast range of water ice landscapes that dominate its surface geology. Its varied terrain suggests a complex climatic and geological history, where water has played a crucial role. Similarly, Charon, Pluto's largest moon, displays extensive resurfacing by water ice, suggesting an early ocean that might have frozen.

Lastly, comets, widely considered as the leftovers from the solar system's formation, have their compositions dominated significantly by water ice. Cometary missions like Rosetta to comet 67P/Churyumov-Gerasimenko have shown not just the presence but also the dynamic behavior of water in these small bodies' sublimation, which feeds the coma and tails seen from Earth.

The distribution of water and ice throughout the solar system underscores its significance in both driving geological processes and sustaining biochemical potentials. Each celestial body tells a part of a larger narrative of solar system history, evolution, and the capacity for hosting life. As we continue to explore space, our understanding of the role of water and ice deepens, ushering a nuanced appreciation of its distribution sources and the intricate dependency of life on this essential molecule wherever it may find a foothold.

5.4 Chemical Reactions Involving Water in Space

Water, with its simple molecular structure of H_2O, remains a cornerstone within the pantheon of molecules crucial for space chemistry. This section delves deeply into the varied chemical reactions that water participates in across different spatial environments, offering poignant insights not only on its reactivity but also its diverse roles in shaping space chemistry and potentially supporting extraterrestrial life.

One of the fundamental reactions involving water in space is its dissociation. Exposed to the harsh radiation environments found in space, particularly ultraviolet (UV) light from the sun, water molecules can undergo photodissociation. This reaction can be represented as:

$$H_2O + h\nu \rightarrow H + OH$$

where $h\nu$ represents the photon of UV radiation. This reaction produces hydroxyl radicals (OH) and hydrogen atoms (H), both of which are extremely reactive. These free radicals can participate in further synthesis processes, forming complex organic compounds on bodies such as comets, asteroids, and planetary surfaces. This mechanism is thought to contribute to the prebiotic chemistry necessary for the origin of life.

On icy moons and comets, another critical set of reactions involves the interaction of water with other volatile elements such as carbon dioxide, methane, and ammonia in the presence of a catalyst or under the influence of radiation. For example, irradiation of ice containing these molecules can lead to the formation of complex molecules like amino acids. A case in point is the experimentally simulated conditions of the icy moons, where irradiation of carbon dioxide and ammonia in water ice results in urea, a molecule central to metabolic processes in living organisms:

$$CO_2 + 2NH_3 + H_2O \rightarrow H_2NCONH_2 + O_2$$

Water also reacts with silicate minerals on dusty or rocky surfaces, initiating hydration or hydroxylation processes that alter mineral structures and could potentially play a role in atmospheric formation or modification around small celestial bodies. This hydration process is vital, especially during planetary formation and in the ongoing evolution of planetary surfaces. In an exogenous context, silicates combining with water create hydrated minerals, evident in Martian geology and extensively studied through Martian rovers.

The role of water extends into the geochemical networks found in the sub-surface oceans of icy moons such as Europa and Enceladus, where it acts not only as a solvent but also as a participant in redox reactions. These reactions might support subsurface microbial life by creating a flux of chemical energy, through processes like serpentinization, where olivine in contact with water and carbon dioxide yields serpentine, magnetite, and methane:

$$3Mg_2SiO_4 + 2H_2O + CO_2 \rightarrow 2Mg_3Si_2O_5(OH)_4 + Mg_3FeSi_2O_5(OH)_4 + CH_4$$

The comprehensive understanding of water's role in space cannot be overstated, from participation in life-supporting reactions to its function as a substratum for various cosmic phenomena. Thus, knowledge of the chemical behaviors of water in extraterrestrial conditions is paramount for predicting not only the possibilities of life elsewhere in the Universe but also understanding essential planetary processes and preparing strategies for future space exploration where water acts as a resource or hazard.

5.5 Ice in Asteroids and Comets: Composition and Significance

Ice, an essential component of many celestial bodies in our Solar System, plays a fundamental role in understanding both the origin of the system and the potential for life beyond Earth. Particularly, ice within asteroids and comets offers invaluable insights into the primordial matter that contributed to the formation of planetary bodies. This section delves into the composition of ice located in asteroids and comets and explores its significance within the context of space chemistry.

Asteroids and comets are often described as the debris left over from the solar system's formation, encapsulating pristine materials from its early history. Many comets, categorized broadly into short-period comets and long-period comets, originate either from the Kuiper Belt or the Oort Cloud, respectively. These regions house bodies with significant ice components, including water ice, carbon dioxide ice, methane ice, and ammonia ice. Notably, the famous comet

5.5. ICE IN ASTEROIDS AND COMETS: COMPOSITION AND SIGNIFICANCE

67P/Churyumov-Gerasimenko, studied meticulously by the Rosetta spacecraft, displayed a diverse range of molecular ices.

The spectroscopic studies performed by the Rosetta's instruments identified not only water ice but also more complex organic molecules embedded within the ice. This molecular diversity suggests that comets could be carriers of complex chemistry essentials for the genesis of life. These findings support the hypothesis that comets, pelting the young Earth, potentially delivered the necessary precursors for prebiotic chemistry.

On the other hand, asteroids, particularly carbon-rich C-class asteroids, have been found to contain traces of water ice. These findings have considerable implications, considering the generally dry surface environments of most asteroids. The presence of water ice, detected through infrared spectroscopy missions dedicated to studying asteroids like Ceres, hints at the possibility of subsurface ice reservoirs. These reservoirs, protected from the sun's heat, could maintain water in solid form and act as time capsules, preserving the chemical conditions predating the formation of the planets.

The study of ice's molecular structure in these celestial bodies—and the distribution of different types of ices—offers clues about the thermal evolution of asteroids and comets. As these bodies travel through the solar system, experiencing varying degrees of thermal stress, their ice content sublimes or changes state, thus providing dynamic insights into their compositional evolution over time.

In comets, the sublimation of ice near the sun results in the familiar cometary coma and tail. This sublimation process, significantly influencing the physical appearance of comets during their perihelion, exposes underlying materials and ejects both dust and gas into space. Thus, analysis of the coma and tail through remote sensing or in-situ sampling provides a detailed chemical profile of the sublimated gases and entrapped solids, enabling a deeper understanding of comet chemistry.

Moreover, the significance of ice in asteroids and comets transcends scientific curiosity and extends into tangible applications. For instance, understanding the distribution and state of ice in comets and asteroids can assist future missions aimed at resource utilization. These celestial ice reservoirs could serve as potential sources of water for human missions, reducing the need to transport water from Earth and supporting longer-term habitation strategies.

Ice in asteroids and comets is a focal point for studying the molecular history and potential habitability in the solar system. The continued exploration and analysis of these icy bodies provide not only critical insights into the fundamental chemical processes that have shaped the cosmos but also offer practical resources for the future of human space exploration. Understanding their composition and leveraging their resource potential stand as pillars in the ongoing journey of space discovery.

5.6 The Role of Water in Supporting Life on Other Planets

Water, with its unique chemical and physical properties, stands as an essential element in the quest for extraterrestrial life. The ubiquitous presence of water across various celestial bodies, particularly on Mars and the icy moons of the outer solar system such as Europa and Enceladus, suggests its integral role in astrobiology. This section explores how water supports the hypothesis of life beyond Earth, detailing its chemical interactions and environmental impacts in extraterrestrial settings.

Water is a vital solvent in biological processes, capable of facilitating chemical reactions necessary for life. In the aqueous environments of other planets and moons, water acts as a medium where organic compounds can dissolve, interact, and possibly lead to the emergence of life. The solvation properties of water enable the transport of ions and molecules, which is critical in cellular functions such as nutrient uptake, waste removal, and energy transfer. These are processes that parallel the biochemical pathways found in terrestrial organisms, suggesting a universal principle for life.

The role of water in thermoregulation provides further evidence for its significance in extraterrestrial environments. Water has a high specific heat capacity, meaning it can absorb or release substantial amounts of heat with only a slight change in temperature. This property could help maintain a stable environment conducive to life on planets with extreme temperature fluctuations. Additionally, water's ability to exist in various physical states—solid, liquid, and vapor—under different environmental conditions contributes to its role in shaping planetary climates. For instance, Earth-like planets with sufficient water could potentially have a hydrological cycle, influencing temperature distribution and climate stabilization through processes like evaporation and condensation.

Detecting subsurface oceans beneath ice-covered moons and the atmospheric water vapor on exoplanets highlights the advanced technological methods used in astrochemical studies. The spectroscopic analysis of light emitted or absorbed by these celestial bodies allows scientists to quantify the molecular signatures of water. Inferences about the presence of liquid water beneath the icy crusts of moons such as Europa are based on gravitational measurements and magnetic field anomalies observed during spacecraft flybys. These findings suggest the existence of hydrothermal vents on these moons, similar to Earth's oceanic vents known to harbor life.

Moreover, water's potential to facilitate life extends to its role in providing protective habitats. Radiation shielding properties of water ice, noted in sections of Martian and lunar polar caps, protect potential biological entities from harmful cosmic and solar radiation. This natural protection mechanism enhances the feasibility of life persisting on planets lacking a robust atmospheric shield.

Water's role in supporting life on other planets is multifaceted, extending from being a vital solvent, a medium for chemical reactions, a stabilizer of climates, to a protector against radiation. Each of these attributes not only highlights the chemical versatility of water but also underscores its universal importance as a cornerstone for life, both on Earth and potentially on other celestial bodies. Understanding these roles transcends scientific curiosity, providing critical insights that fuel the ongoing exploration of our universe in search of life beyond our planet.

5.7 Techniques for Extracting and Analyzing Water and Ice in Space

The extraction and analysis of water and ice from celestial bodies are critical to advancing our understanding of space environments and the sustainability of life beyond Earth. Techniques for extracting and analyzing these substances utilitze cutting-edge technology and sophisticated methodologies that reflect the unique challenges posed by extraterrestrial environments.

Extraction Techniques: Extraction of water and ice in space primarily targets celestial bodies such as the Moon, Mars, and icy moons and comets. Different methodologies are adapted based on the physical state and location of the water ice.

Sublimation Extraction: Utilizing the direct transition of ice to vapor, sublimation units heat the subsurface ice, bypassing the liquid phase. This vapor is then captured and condensed back into liquid water in a closed system. This method is particularly useful in vacuum conditions where the boiling point of water is significantly lower.

Electrolytic Melting: This technique involves the application of direct current to heat the ice through resistance heating, melting it into a liquid which can then be extracted. This method is useful for ice embedded deep within rocks or soil and can be efficiently carried out with minimal thermal loss to the environment.

Robotic Drilling: Robotic drills are deployed to penetrate the surface of celestial bodies to reach subsurface ice reservoirs. The extracted cores are then heated, and the melted water is pumped out. Advances in autonomous robotics enhance the efficiency and depth of penetration possible with this technique.

Analytical Techniques: Once extracted, the water and ice undergo thorough chemical analysis to assess their composition, quality, and potential for supporting life.

Spectroscopy: Both infrared and Raman spectroscopy are utilized extensively to determine the molecular composition of ice and water samples. These techniques provide essential data on the presence of organic compounds and possible contaminants.

Mass Spectrometry: Utilized to analyze the isotopic ratios of hydrogen and oxygen in water, mass spectrometry helps in understanding the origin and history of water molecules in space. It also provides data on the presence of other volatile elements and compounds within the ice.

Chromatography: Gas chromatography and liquid chromatography are employed to separate and identify compounds dissolved in the water. This method is vital for assessing the potential nutritive value or toxicity of the water in support of human activities or local biotic systems.

Challenges and Innovations: The extreme conditions of space environments pose significant challenges to the extraction and analysis of water and ice. These challenges include temperature extremes, radiation, microgravity conditions, and the vacuum of space. Innovative solutions, such as the development of cryogenic drilling techniques and the implementation of autonomous operations, have progressively overcome many of these challenges.

Moreover, the ongoing miniaturization and enhancement of analytical instruments have allowed for more efficient, accurate, and detailed analyses of extracted samples. The integration of AI and machine learning into data analysis processes further enhances the prediction and modeling capabilities of scientists working in this field.

In summary, the techniques for extracting and analyzing water and ice in space are characterized by a blend of advanced engineering, robotic technology, and sophisticated analytical methods. These efforts are not only pivotal for the sustained presence of humans in space but also provide critical data for astrobiological studies, enhancing our understanding of the possibilities for life in the universe.

5.8 Water as a Shielding Material against Space Radiation

Water, despite its simplicity, is a multifaceted molecule that plays varied and vital roles in the context of space exploration. Notably, its utility transcends merely being a sustainer of life or a potential fuel source; water also emerges as a candidate for radiation shielding - a critical consideration in the protection of astronauts from harmful cosmic rays and solar radiation.

Radiation in space presents a formidable challenge to manned and unmanned space missions. It includes a spectrum of particles such as protons, alpha particles, and heavy ions, along with electromagnetic radiation like X-rays and gamma rays, all of which are known for their potential to cause biological damage and interfere with spacecraft operational integrity. Traditional shielding materials used on Earth, such as lead or concrete, are significantly less practical for space due to their high mass and associated launch costs.

The physical properties of water, namely its high hydrogen content, make it a prime candidate for radiation shielding. Hydrogen-rich materials are effective at attenuating fast-moving charged particles because these particles lose energy upon collisions with light nuclei, primarily through ionization processes. Water, with its high proportion of hydrogen atoms, can effectively slow down and stop these particles, reducing their damaging effects. Moreover, water has a benign nature and does not produce secondary radiation that can be as harmful as the initial gamma rays or neutrons, a phenomenon often observed with heavier elements like lead.

In-depth analysis and experiments conducted in simulated space environments suggest that water can reduce the dose equivalent of both primary and secondary radiations to levels that are compatible with NASA safety standards. For example, experimental data indicated that a water layer approximately 40 cm thick could reduce the exposure to high-energy protons and heavy ions from solar and cosmic sources to acceptable mission levels.

The practical application of water as a radiation shield involves innovative spacecraft design strategies. For instance, the walls of habitable modules in spacecraft could be lined or filled with liquid water tanks, or water-embedded materials could be used. Additionally, the dual-usage of onboard water for both life support and radiation protection optimizes the available mass and space, an ongoing challenge in spacecraft design.

Further underscoring its suitability, water's inevitable centrality in future lunar and Martian bases (for consumption, agriculture, and possibly fuel through hydrolysis) reinforces its role as a shielding material. Proposed architectural ideas for extraterrestrial habitats include using water ice combined with regolith to shield from both radiation and extreme temperature variations. This approach not only provides efficient radiation shielding but also utilizes in-situ resources, thus aligning perfectly with the concept of sustainability in space exploration.

From a theoretical perspective, understanding the shielding properties of water also involves complex calculations of energy loss per unit path length (linear energy transfer) in water vs. other potential materials. These calculations take into account the specific radiation types encountered in space and the spectrum of energies involved, ensuring that space mission planners can accurately gauge and implement sufficient protection.

Conclusively, as we venture further into space, the importance of developing multi-functional and efficient materials like water becomes increasingly clear. Water is not only vital to sustaining life but also plays an essential role in ensuring the safety and success of future space missions through its capacity to act as an effective radiation shield.

Problems

1. Calculate the minimum thickness of a water shield required to reduce the radiation from a solar proton event to safe levels for a crewed mission to Mars.

2. Discuss the potential design considerations for incorporating water-based radiation shielding into a spacecraft, considering both functionality and the impact on overall spacecraft mass.

3. Evaluate the effectiveness of combining regolith and water ice for use in constructing an extraterrestrial habitat. Consider factors such as material availability, environmental conditions, and expected radiation types.

4. Using the concept of linear energy transfer, explain why hydrogen-rich substances like water are more effective at shielding against certain types of space radiation compared to heavier materials like lead.

5. Design an experiment to test the efficiency of water as a radiation shield in low Earth orbit, outlining the methodology, expected challenges, and measurements to be taken.

5.9 Utilization of Water and Ice for Fuel and Life Support Systems

Water and ice are not only critical substances for supporting life through hydration and as a component of biological systems, but they also play a pivotal role in the functionality of space habitats through their utilization in life support systems and as a source for fuel production. This section elucidates the dual application of water and ice in these capacities, highlighting the technical methodologies and the underlying chemical principles governing their use in space exploration.

In space environments, water is indispensable for life support systems. It is used for drinking, food preparation, hygiene, and environmental control within habitats. Recycling and purifying water onboard spacecraft and space stations is therefore a cornerstone of life support strategies. The closed-loop water recovery systems employed mimic Earth's natural water cycle, involving processes such as filtration, bioreaction, distillation, and condensation. An example of such a system is the Water Recovery System (WRS) used on the International Space Station (ISS), which recycles about 90% of the water onboard. Water molecules are split into hydrogen and oxygen through electrolysis; the oxygen is then used for breathing air, while the hydrogen is vented out or recycled.

Furthermore, ice, particularly as found on extraterrestrial bodies such as the Moon or Mars, presents a valuable resource for in situ resource utilization (ISRU). Robotic missions to the lunar poles, for example, are primarily motivated by the potential to mine ice deposits, which could be used to support human colonies or produce rocket fuel. The presence of regolith layers over these ice deposits also raises interesting chemical challenges and opportunities. The layers protect the ice but require innovative mining techniques to access the water without contaminant introduction. Once accessed, the harvested ice could be melted, purified, and either used directly or split into hydrogen and oxygen via electrolysis.

The splitting of water into hydrogen and oxygen in a microgravity environment is facilitated by the use of onboard electrolytic cells. This practice not only supports life by providing oxygen but also fuels future exploration missions. Hydrogen, when combined with oxygen, can serve as rocket propellant, opening avenues for long-duration space missions and the exploration of farther celestial bodies. This elemental release and recombination process—occurring through a unit reaction in the cell known as a proton exchange membrane (PEM)—exemplifies a practical application of chemical principles in an extraterrestrial context.

The twin utility of water in life support and as a source of fuel necessitates highly efficient systems designed to maximize resource availability while minimizing waste. The challenges therein lie not only in the technical execution of such systems but also in their optimization and adaptation to different environments and mission parameters. For instance, the gravity variances between Earth, the Moon, and Mars affect how water behaves and is managed, thereby influencing system design and functionality.

The potential of using water as a shielding material against radiation additionally complements its utility, creating a multifunctional resource that underscores the critical importance of managing and understanding this substance in space exploration. The ability for water to serve multiple critical functions including human sustenance, habitat construction, and propulsion highlights its indispensable role in space exploration logistics.

Problems for consideration:

- Design a model for a closed-loop water recovery system suitable for use in a Martian habitat. Consider factors such as atmospheric composition, gravity, and temperature.

- Calculate the amount of energy required to split 1000 kg of water into hydrogen and oxygen using electrolysis. Discuss how differing gravitational fields might impact the energy efficiency of this process.

- Propose a method for mining ice buried beneath lunar regolith that minimizes contamination and maximizes efficiency. Consider technological, environmental, and chemical safety factors.

5.10 Challenges in the Management and Recycling of Water in Space Habitats

Water management in space habitats presents a series of intricate challenges crucial for the sustainment of life and the successful operation of missions. The closed environment of a space habitat necessitates a nearly 100% recycling rate of water, and achieving this high level of efficiency encompasses numerous technological, microbiological, and logistical hurdles.

One of the foremost technical challenges is the development and maintenance of an efficient, robust water recovery system capable of handling the diverse use of water in a space habitat. Water onboard is used not just for drinking, but also for food preparation, hygiene, and cooling of equipment. Each use introduces different contaminants into the water supply, ranging from organic compounds and microorganisms from human waste to inorganic salts and trace chemicals from machinery. The water recovery systems must effectively remove these contaminants to prevent the accumulation of harmful substances and deterioration of water quality over time.

The primary methods currently employed for water purification in space include distillation, filtration, and catalytic oxidation. Each method has its limitations and efficiency constraints. For example, distillation requires significant energy inputs and careful temperature control to be effective, while filtration systems necessitate regular replacement of filters that can accumulate and harbor pathogenic bacteria. Moreover, the reduced gravity conditions of space impact fluid dynamics, complicating the separation processes based on settling or phase separations, typically used on Earth.

Microbial contamination poses another substantial challenge. The closed environment of a space habitat provides a unique and potentially conducive environment for the growth and mutation of microorganisms. Controlling these populations is critical, as unchecked growth can lead to biofilm formation, which can corrode and clog water processing equipment and pose health risks to the crew. Existing biocides or antimicrobial agents must be used judiciously to avoid creating resistant strains of microorganisms while ensuring they do not introduce harmful by-products into the water supply.

Logistical issues also play a significant role in the management of water in space habitats. The limited volume and mass capacity of space missions necessitate careful consideration of the balance between water supply, recycling equipment, and other essential supplies and equipment. The installation of comprehensive water recovery systems must not compromise other vital systems due to space or weight constraints.

The integration of advanced water recovery systems also requires significant energy inputs, a precious commodity in space. Energy-efficient technologies thus become paramount, but these must be balanced against the system's overall efficacy and the potential need for increased power supply, which itself could influence mission logistics and costs.

In response to these challenges, ongoing research and development are focused on advancing water recovery technologies, optimizing energy use, and enhancing system reliability and lifespan. Innovative approaches such as the use of nanotechnology for filtration and the incorporation of artificial intelligence for monitoring and managing water quality and distribution systems are among the avenues being explored.

To further address these challenges, simulating space habitat conditions for extended periods on Earth allows for the testing and refinement of water recycling technologies. Such simulations help in understanding the behavior of water recovery systems over long durations, providing invaluable data that can inform design improvements.

Despite the complexities involved, the effective management and recycling of water in space habitats are achievable, with continuous advancements in technology, understanding of microbiological behaviors in closed environments, and thorough planning and testing. The pursuit of greater efficiency and reliability in these systems remains critical as humanity aims toward longer-duration space missions and the eventual habitation of other planetary bodies.

Chapter 6

Atmospheres of Planets and Moons: Chemical Analysis

This chapter focuses on the chemical composition and analysis of the atmospheres of various planets and moons. It outlines the techniques used to investigate atmospheric gases, including spectroscopy and mass spectrometry, and discusses the chemical processes that occur within these gaseous envelopes. Key subjects include the diverse atmospheres of terrestrial planets, gas giants, and icy moons, highlighting how their chemical makeups can provide insights into climatic conditions, geological history, and potential for supporting life.

6.1 Introduction to Planetary Atmospheres

The exploration of planetary atmospheres begins with understanding their fundamental properties and the significance they hold in the characterization of celestial bodies within our solar system and beyond. By examining the composition, structure, and dynamic behaviors of these atmospheres, scientists can infer valuable insights about the climatological and geological histories of planets and moons, as well as assess their potential to support life.

CHAPTER 6. ATMOSPHERES OF PLANETS AND MOONS: CHEMICAL ANALYSIS

A planetary atmosphere is essentially a layer of gases surrounding a planet or moon, held in place by the body's gravity. The complexity and composition of these atmospheres vary significantly from one celestial body to another, influenced by factors such as mass, temperature, radiation levels, and volcanic activity. Each atmosphere presents a unique combination of gases, often dominated by one or a few primary constituents, alongside trace amounts of a wide variety of other compounds.

The Earth's atmosphere, predominantly composed of nitrogen and oxygen, stands as an illustrative model of a life-supporting system, enabling a comparative basis for studying more exotic atmospheres encountered elsewhere. Contrasting sharply with Earth, the toxic and dense carbon dioxide-dominated atmosphere of Venus provides a stark example of a runaway greenhouse effect, while the thin, carbon dioxide-rich atmosphere of Mars gives clues about possible planetary climate changes. Furthermore, the predominantly hydrogen and helium atmospheres found on gas giants like Jupiter and Saturn, and the complex organic chemistry in the atmospheres of icy moons such as Titan, extend the diversity of atmospheric compositions.

Atmospheric pressure and temperature are key to determining the state and behavior of the substances in an atmosphere. These environmental conditions affect cloud formation, weather patterns, and climatic cycles, which in turn influence the surface and subsurface conditions of the celestial body. For instance, understanding Jupiter's atmospheric pressure and temperature gradients is vital to studying its deep cloud formations and the complex interplay of its atmospheric components.

Chemical processes occurring in these atmospheres, such as photochemical reactions driven by solar ultraviolet radiation, play crucial roles in shaping atmospheric characteristics. On Titan, for example, these processes lead to the production of complex hydrocarbons and nitriles, generating a thick haze that obscures its surface.

The study of planetary atmospheres is not only about understanding their current state but also about unlocking the secrets of planetary evolution and the potential habitability of other worlds. This requires sophisticated analytical techniques and robust theoretical models to interpret observational data. As such, the introduction to planetary atmospheres serves as the cornerstone of astrochemical researches, providing a foundational understanding that supports further in-depth studies into the unique atmospheric phenomena exhibited by different celestial bodies.

The study of planetary atmospheres offers profound insights into the physical and chemical processes that govern the evolution of planets and moons. As we probe these enigmatic layers of gases, we continue to enrich our understanding of the universe, advancing both scientific knowledge and technological frontiers in our quest to explore the cosmos.

6.2 Fundamentals of Atmospheric Chemistry

Understanding the basic principles of atmospheric chemistry is essential for analyzing the chemical makeup of planetary atmospheres. This section delves into the foundational chemical processes that define atmospheric behavior, including but not limited to photochemical reactions, thermal equilibrium, and catalytic cycles.

At the core of atmospheric chemistry are the principles of thermodynamics and kinetics, which govern the behavior and interaction of molecules in the gaseous state. The atmosphere is primarily a mixture of nitrogen, oxygen, argon, carbon dioxide, and water vapor, but each planetary atmosphere features its own unique composition, influenced by its chemical reactions and physical conditions.

One pivotal aspect of atmospheric chemistry is the concept of photochemistry, which involves chemical reactions initiated by light. In planetary atmospheres, ultraviolet (UV) radiation from the parent star can break chemical bonds, leading to the creation of highly reactive radicals. For instance, in Earth's upper atmosphere, UV radiation dissociates diatomic oxygen (O_2) into two oxygen atoms (O), which can then react with other O_2 molecules to form ozone (O_3). This reaction is crucial for life, as ozone absorbs and considerably reduces the amount of harmful UV radiation reaching the surface.

Ozone formation and destruction illustrate a dynamic photochemical equilibrium, which is vital in atmospheric chemistry. This balance can be disturbed by other chemical species acting as catalysts. For example, chlorofluorocarbons (CFCs) release chlorine atoms in the stratosphere, which catalyze the decomposition of ozone and lead to ozone layer depletion.

Another fundamental principle is the role of chemical kinetics, which describes the rates of chemical processes. Atmospheric reactions can vary dramatically in their time scales, from the very fast (fractions of a second) to the extremely slow (decades). Understanding these rates is crucial for predicting atmospheric behavior, particularly in response to sudden changes such as volcanic eruptions or human-made pollution.

Atmospheric pressure and temperature also play critical roles in determining the rate and extent of chemical reactions. For example, the decrease in temperature with altitude in Earth's atmosphere means that many reactions that are unfeasible at ground level (due to kinetic energy barriers) can occur at high altitudes.

In planetary atmospheres, such as those of Venus or Jupiter, extreme conditions can lead to the formation of exotic chemical compounds and behavior. Venus's surface pressure and temperature are so high that they lead to a supercritical state where the carbon dioxide-rich atmosphere exhibits properties of both gases and liquids. Meanwhile, Jupiter's upper atmosphere contains hydrogen, helium, and methane, the interactions of which under high pressure and radiation fields result in complex organic molecules, which are entirely different from those seen on Earth.

Understanding these chemical fundamentals allows for deeper insights into the atmospheres of different planets, prediction of environmental conditions, and even the potential habitability of extraterrestrial bodies. This forms the bedrock upon which further detailed analytical techniques, as discussed later in this textbook, are applied to explore and comprehend the vast chemical systems present in planetary atmospheres.

Problems

To reinforce the concepts discussed, consider the following problems:

- Calculate the potential energy change for the formation of ozone from oxygen under standard conditions.
- Discuss the role of catalytic cycles in the atmospheric breakdown of methane in Jupiter's atmosphere.
- Design a simple model using chemical kinetics to predict the formation of nitrogen oxides from lightning strikes in Earth's atmosphere.

- Analyze how changing atmospheric pressure could theoretically affect the chemical makeup of Saturn's atmosphere.

These problems require a combination of analytical skills and theoretical knowledge, challenging students to apply their understanding in practical settings.

6.3 Common Gases in Various Planetary Atmospheres

As we advance in our exploration of planetary atmospheres within our solar system, the understanding of their chemical composition becomes imperative. The atmospheres of different planets and moons are marked by distinct gases, which not only reflect the physical and geological processes that occur on these celestial bodies but also their evolutionary histories.

In the terrestrial planets, such as Earth, Venus, and Mars, the atmospheres are primarily composed of carbon dioxide (CO_2), nitrogen (N_2), and, in Earth's case, oxygen (O_2). Venus's atmosphere, with over 96% carbon dioxide, reflects its runaway greenhouse effect, whereas Mars, with a thin atmosphere containing about 95% carbon dioxide, reveals both its geological activity and the impacts of solar wind stripping. Earth's atmosphere is unique among terrestrial planets due to its substantial oxygen content, a result of biological processes through photosynthesis. This diversity illustrates the influence of both geological and biological processes on atmospheric composition.

Moving further away from the Sun, the giant planets—Jupiter, Saturn, Uranus, and Neptune—exhibit atmospheres dominated by hydrogen (H_2) and helium (He), elements that were most abundant in the protoplanetary nebula during the formation of the solar system. The presence of methane (CH_4), ammonia (NH_3), and water vapor (H_2O) in smaller amounts further demonstrates the low-temperature condensation processes that affect heavier molecules. These gases, combined with traces of hydrogen sulfide (H_2S) and phosphine (PH_3), provide critical clues about the thermodynamic conditions and the chemical equilibrium within these intense and high-pressure atmospheres.

Titan, Saturn's largest moon, presents a fascinating case with an atmosphere rich in nitrogen, similar to Earth, but with methane as the second-most abundant component. The photochemical reactions triggered by solar ultraviolet radiation lead to the formation of complex organic molecules such as ethane (C_2H_6), diacetylene (C_4H_2), and acetylene (C_2H_2), making Titan a prime candidate for studying prebiotic chemistry.

According to spectroscopic analysis, the thin atmospheres of several moons including Europa and Enceladus, which are primarily composed of oxygen and water vapor respectively, suggest the outgassing from their icy surfaces or subsurface oceans. This phenomenon highlights the dynamic interaction between the surface and the atmosphere in maintaining the chemical equilibrium and facilitating the transport of organic molecules.

The diversity in atmospheric compositions across different planetary bodies provides indispensable insights into their origins and current states. It is crucial for understanding the potential habitability and the presence of life beyond Earth. Utilizing techniques such as spectroscopy and mass spectrometry for their analysis permits a detailed understanding of the atmospheres' chemical makeup, thus opening avenues for future explorations aimed at uncovering the mysteries of our solar system.

The variability in the atmospheric compositions amongst planets and moons significantly enhances our understanding of planetary science. It lays a solid foundation for both theoretical studies and experimental simulations, vital for the continuous exploration and eventual colonization of these celestial bodies.

6.4 Analytical Techniques for Atmospheric Analysis

The detailed analysis of planetary atmospheres is crucial for understanding their composition, climate, and potential for supporting life. A variety of sophisticated techniques are employed to discern and quantify the types and concentrations of gases present. These methods include, but are not limited to, spectroscopy, mass spectrometry, and chromatography. Each technique offers unique insights into the molecular and elemental signatures of an atmosphere and is chosen based on the specific objectives of the study.

6.4. ANALYTICAL TECHNIQUES FOR ATMOSPHERIC ANALYSIS

Spectroscopy is perhaps the most widespread technique used in the study of planetary atmospheres. This method leverages the interaction of light with matter to detect and quantify various gases. When light of specific wavelengths is absorbed, emitted, or scattered by atmospheric particles, it can provide a fingerprint unique to each chemical substance. There are several types of spectroscopy, including infrared (IR), ultraviolet (UV), visible (VIS), and X-ray, each suited to different ranges of the electromagnetic spectrum and thus sensitive to various atmospheric components. For instance, infrared spectroscopy is particularly effective in detecting molecular vibrations and rotations, making it invaluable for identifying greenhouse gases like carbon dioxide and methane.

Mass spectrometry provides another critical set of tools for atmospheric analysis. This technique measures the mass-to-charge ratio of ionized atoms or molecules to determine their composition. In planetary science, mass spectrometers are often deployed on orbital and landed missions. They allow for the direct sampling of an atmosphere or the analysis of surface outgassing, delivering precise measurements of isotopic and elemental abundances. A notable application is on the Mars Curiosity Rover, which uses a mass spectrometer to analyze air and rock samples to infer the planet's past habitability.

Chromatography, particularly gas chromatography, is used to separate and analyze volatile compounds in an atmospheric sample. When combined with mass spectrometry (GC-MS), this method can provide detailed chemical profiles that are critical for understanding complex organic molecules found within an atmosphere. It is especially useful in the analysis of Titan's thick, nitrogen-rich atmosphere, where complex hydrocarbons and other organic substances abound.

All these techniques rely heavily on data interpretation methods and calibration standards that stem from terrestrial environments but must often be adapted for extraterrestrial conditions. For instance, spectroscopic data from distant planets must be carefully corrected for distortive effects caused by the interstellar medium and the instrument's own environment. This introduces a layer of complexity in data analysis, necessitating sophisticated algorithms and modeling techniques often tailored to specific missions.

Specialized instruments, such as the *Atmospheric Chemistry Suite* (ACS) and *Nadir and Occultation for Mars Discovery* (NOMAD) on the ExoMars Trace Gas Orbiter, exemplify advancements in the integration of multiple analytical techniques like infrared and ultraviolet spectroscopy to provide a holistic view of Martian atmospheric dynamics.

These analytical techniques not only enhance our understanding of the composition and behavior of planetary atmospheres but also continuously refine our ability to detect signs of extraterrestrial life and assess planetary habitability. They illuminate chemical processes such as photochemical reactions and atmospheric escape mechanisms, adding depth to our knowledge of planetary science.

In summary, the precision and adaptation of analytical techniques like spectroscopy, mass spectrometry, and chromatography are fundamental in advancing our exploration and understanding of planetary atmospheres. As our technological capabilities expand, the future of atmospheric analysis looks promising, set to unravel more complexities of our solar system and beyond.

6.5 Venus: Atmospheric Composition and Chemical Phenomena

Venus, often referred to as Earth's sister planet due to its similar size and proximity, presents a stark contrast in atmospheric composition and phenomena. Dominated by a thick, toxic atmosphere primarily composed of carbon dioxide (CO_2) with traces of nitrogen (N_2), sulfur dioxide (SO_2), and water vapor (H_2O), Venus offers a unique environment for the study of chemical and climatic extremes.

The dense Venusian atmosphere, with a pressure about 92 times that of Earth's at sea level, hosts an average temperature of approximately 737 K (464 °C; 867 °F). This high temperature is primarily the result of a potent greenhouse effect caused by the massive concentration of CO_2. The CO_2 absorbs infrared radiation from the planet's surface, trapping heat and raising the overall temperature of the atmosphere.

Sulfur compounds play a significant role in Venus's atmospheric chemistry. Clouds of sulfuric acid (H_2SO_4) droplets are suspended in the thick carbon dioxide layer. These clouds are formed from sulfur dioxide (SO_2), which, under solar ultraviolet (UV) radiation, undergoes a series of photochemical reactions. The SO_2 is oxidized, primarily to SO_3, which then reacts with water vapor to form sulfuric acid. This conversion is facilitated by the presence of atmospheric catalysts, likely chloride and fluoride compounds.

6.5. VENUS: ATMOSPHERIC COMPOSITION AND CHEMICAL PHENOMENA

The presence of trace amounts of hydrochloric acid (HCl) and hydrofluoric acid (HF) in the atmosphere also indicates volcanic activity, which releases these gases into the atmosphere. These halides, in conjunction with sulfur dioxide and water vapor, contribute to a complex cycle of chemical reactions that result in the thick, opaque clouds that obscure the planet's surface from direct observation.

Lightning and thermal gradients also contribute to the dynamic atmospheric phenomena on Venus. Electrical discharges within the clouds can influence chemical reactions, potentially synthesizing new compounds. These electrical phenomena are thought to be catalysts for the formation of trace organic molecules detected in the atmosphere.

One of the most intriguing aspects of Venusian chemistry is the hypothesized presence of phosphine (PH_3). Phosphine, a phosphorus and hydrogen compound typically associated with the decay of organic matter on Earth, was tentatively detected in trace amounts in the Venusian clouds. This discovery has sparked significant interest because on Earth, phosphine is primarily produced through biological processes. The potential biotic or abiotic production of PH_3 on Venus is a subject of ongoing research and debate, highlighting the possibility of unknown chemical pathways or even suggestive of past or present life.

Spectroscopic techniques, particularly those utilizing ultraviolet, infrared, and radio wavelengths, have been instrumental in unveiling the composition and behavior of the Venusian atmosphere. These methods allow scientists to identify specific gases, study their distribution and concentration, and observe ongoing chemical reactions from afar. Future missions to Venus may carry advanced spectroscopic and sampling instruments aimed at conducting in-situ analysis, providing deeper insights into its atmospheric chemistry.

The extreme conditions and bizarre chemical phenomena observed on Venus provide a valuable comparative framework for understanding planetary atmospheres. Investigations into its hostile yet fascinating atmospheric environment not only broaden our knowledge of planetary science but also enhance our understanding of Earth's own atmospheric processes and future.

Problems for Consideration:

- Calculate the partial pressure of carbon dioxide in Venus's atmosphere using the ideal gas law, given that its concentration is about 96.5% and the total surface pressure is about 92 atmospheres.

- Using the information provided on sulfuric acid cloud formation, outline the sequence of chemical reactions starting from SO_2 under Venusian conditions.

- Discuss the potential origins of phosphine in Venus's atmosphere and its implications for our understanding of Venusian atmospheric chemistry.

6.6 Mars: Searching for Signs of Life through Atmospheric Chemistry

The exploration of Mars, driven by the quest to find signs of past or present life, relies heavily on understanding its atmospheric chemistry. Mars' atmosphere, primarily composed of carbon dioxide (CO_2) at about 95.3%, also contains nitrogen (N_2) and argon (Ar) as minor components, along with trace amounts of oxygen (O_2), water vapor (H_2O), and methane (CH_4). The presence and variability of methane, in particular, have drawn significant attention due to its potential biological origins.

Methane's temporal and spatial variation on Mars, detected by various missions like the Mars Curiosity Rover and orbital spectrometers, proposes a dynamic process at work. On Earth, methane is largely produced by biological sources, but it can also arise from abiotic processes such as serpentinization, which involves the reaction of water with olivine, leading to the formation of serpentine, hydrogen, and potentially methane. The detection of methane spikes on Mars raises the crucial question of their origin - biological or geophysical.

The characterization of Martian atmospheric chemistry is enhanced by spectroscopy, particularly through infrared spectroscopy. This technique allows for the identification of chemical bonds by the absorption patterns each compound creates in the infrared region of the electromagnetic spectrum. Tools like the Tunable Laser Spectrometer (TLS) aboard the Curiosity Rover have facilitated detailed methane measurements, providing data that suggests episodic increases that may be linked to seasonal temperature changes.

Further, isotopic ratios in the Martian atmosphere present another avenue through which chemical processes can be understood. Isotope studies, particularly involving carbon and oxygen, help distinguish between biological and geological methane sources. Biological processes typically show preference for lighter isotopes, leading to distinct isotopic signatures compared to abiotic processes.

The role of ultraviolet (UV) radiation must also be considered in the atmospheric chemistry of Mars. UV exposure can lead to the production of reactive species from otherwise stable atmospheric gases, influencing not only the potential for supporting life but also the stability of biological molecules. Reactive species generated from CO_2 and H_2O under UV light include hydroxyl radicals ($OH\cdot$), which affect methane's stability, providing a possible sink for atmospheric methane independent of biological activity.

Moreover, Mars' dusty environment contributes uniquely to its atmospheric chemistry. Dust storms can lift water vapor and other gases to higher altitudes, increasing their exposure to UV radiation and possibly leading to chemical reactions that would not occur at lower altitudes.

Understanding these complex chemical interactions helps clarify the Martian atmospheric environment's potential habitability. The ongoing analysis of gas interactions, UV radiation effects, and seasonal changes provides essential clues in the quest for life. By integrating these chemical signatures with geological and climatic data, scientists aim to construct a comprehensive picture of whether Mars could have supported microbial life and whether it might still do so.

To probe deeper into the potential biological implications of Martian atmospheric chemistry, further missions will need to focus not just on detecting present gases but also on their interaction dynamics and temporal variations. These studies will go hand in hand with continuous advancements in analytical techniques and modeling approaches, providing a deeper insight into the Red Planet's enigmatic atmosphere.

Evaluation problems:

- Discuss the implications of finding methane in the Martian atmosphere concerning abiotic and biotic sources.

- Derive the reaction pathways for methane production through serpentinization.

- Propose an experimental design to test the impact of Martian dust storms on atmospheric chemical stability using UV radiation.

- Analyze the potential effects of isotopic fractionation in differentiating between biological and geological sources of Martian atmospheric gases.

6.7 Giant Planets: The Role of Hydrogen and Helium in Atmospheric Chemistry

The chemical atmospheres of giant planets—Jupiter, Saturn, Uranus, and Neptune—offer a fascinating glimpse into the complex dynamics of planetary science. These planets, predominantly composed of hydrogen and helium, present an environment where conventional atmospheric chemistry markedly differs due to their unique compositions and physical conditions.

Hydrogen, the lightest and most abundant element in the universe, dominates the atmospheres of these gas giants. Its molecular form (H_2) serves as the primary component of their atmospheres. Helium, the second lightest and second most abundant element, comprises a smaller but significant percentage. The ratio of hydrogen to helium affects many of the thermal, chemical, and dynamic properties of the atmospheres.

The presence of minor volatile compounds like methane (CH_4), ammonia (NH_3), and water vapor (H_2O) is also noteworthy. However, the primary focus remains on hydrogen and helium due to their roles in influencing atmospheric characteristics ranging from heat distribution to magnetic field interactions.

Thermodynamic Properties

In the deep atmospheres of giant planets, the pressure and temperature conditions are extreme, allowing hydrogen to exist in both molecular and metallic forms. This metallic hydrogen is crucial in generating the substantial magnetic fields observed around these planets, like Jupiter's intense magnetosphere. The shift from molecular to metallic hydrogen involves electrons moving freely, similar to the properties observed in metals.

6.7. GIANT PLANETS: THE ROLE OF HYDROGEN AND HELIUM IN ATMOSPHERIC CHEMISTRY

Helium, meanwhile, does not undergo any phase change but remains in a gaseous form, imparting stability and moderating the reactive environment created by hydrogen. The solubility of helium in metallic hydrogen decreases with increasing pressure and temperature—a phenomenon that leads to "helium rain," an exotic form of precipitation theorized to occur within Jupiter and Saturn.

Atmospheric Dynamics

Hydrogen's low molecular mass contributes significantly to the high scale height (vertical change rate in atmospheric pressure) of these planets' atmospheres, defining their extended and diffuse nature. Helium, while denser than hydrogen, complements this by adding to the overall mass without drastically affecting the scale height.

Chemical Interactions

The chemistry of hydrogen and helium alone is fairly limited due to their inert qualities. However, their interaction with minor components like methane leads to fascinating chemical pathways. For example, under ultraviolet irradiation or in the presence of electrostatic discharges (such as lightning), methane can decompose and then reform into complex hydrocarbons such as ethane (C_2H_6) and acetylene (C_2H_2), contributing to the colored bands noticeable in the atmospheres of Jupiter and Saturn.

Photochemical reactions driven by solar radiation play a significant role in these processes. The deeper layers of the atmospheres, shielded from direct solar irradiation, are less chemically active but crucial in the redistribution of heat and the formation of deeper cloud structures.

Implications for Planetary Formation and Evolution

Understanding the role of hydrogen and helium in these massive planetary atmospheres provides insights into the processes of planetary formation and evolution. For instance, the differentiation of helium from hydrogen in the interiors of Jupiter and Saturn can inform us about the thermodynamic histories of these planets. Additionally, this differentiation has implications for their cooling rates and subsequent atmospheric and magnetic field developments.

In sum, the atmospheres of the giant planets serve as natural laboratories for studying fundamental principles of chemistry and physics under conditions unattainable on Earth. The roles of hydrogen and helium, though central, are influenced by minor constituents which further enrich the complexity of atmospheric dynamics and chemistry. As such, giant planets not only add to our understanding of planetary atmospheres but also challenge and expand the known boundaries of chemical science.

6.8 Titan and Other Moons: Complex Organic Chemistry in the Atmosphere

Saturn's moon Titan stands out among the solar system's moons for its dense, chemically rich atmosphere, which is primarily composed of nitrogen and methane, along with a host of other hydrocarbons and organic molecules. This atmospheric composition provides a unique laboratory for studying prebiotic chemistry and the pathways that might lead to life, constituting a prime example of complex organic chemistry in extraterrestrial environments.

To understand the chemistry of Titan's atmosphere, it is essential to consider the role of photochemical reactions driven by solar ultraviolet radiation and electron bombardments from Saturn's magnetosphere. These energy sources initiate complex chains of reactions in the upper atmosphere, leading to the formation of larger molecules and eventually aerosols that settle as organic rain onto Titan's surface. This organic cycle resembles the hypothesized processes on early Earth and offers insights into potential astrobiological processes.

Methane (CH_4) and ethane (C_2H_6) play critical roles in Titan's atmospheric chemistry. Methane is broken down by sunlight to form ethane, acetylene (C_2H_2), ethylene (C_2H_4), and a variety of other hydrocarbons. These reactions contribute to the formation of a thick haze that obscures the moon's surface at visible wavelengths. This haze layer is a rich matrix of organic compounds, primarily consisting of tholins, complex organic polymers formed through the irradiation of simple hydrocarbons.

Amidst this complexity, nitrogen compounds also form, such as hydrogen cyanide (HCN), cyanoacetylene (HC_3N), and others, adding to the rich chemical tapestry of Titan's atmosphere. HCN, a particularly interesting molecule due to its potential role in the formation of

6.8. TITAN AND OTHER MOONS: COMPLEX ORGANIC CHEMISTRY IN THE ATMOSPHERE

amino acids, illustrates how inorganic molecules may transition into complex organic species that are significant for life as we perceive it.

In addition to Titan, other moons in the solar system also display intriguing chemical phenomena in their atmospheres. For instance, Europa, a moon of Jupiter, possesses a tenuous atmosphere composed primarily of oxygen, which is believed to be generated by the radiolysis of water ice on the moon's surface by Jupiter's intense magnetic field. Though not as chemically complex as Titan's atmosphere, the processes at Europa underscore the reactivity and transformation of simple molecules under space weathering conditions.

Ice giants' moons such as Triton, a moon of Neptune, present another interesting case. Triton's thin atmosphere, mostly nitrogen with small amounts of methane, is subject to extreme conditions that contribute to the creation of nitrogen ice and possibly organic compounds via processes similar to those on Titan.

The understanding of these moons' atmospheres is principally obtained through spectroscopic methods from orbiters, telescopes, and in the case of Titan, data from the Huygens probe which landed in 2005. The molecular signatures captured in these datasets provide critical clues on the atmospheric composition and dynamics, offering snapshots of the ongoing chemical transformations.

Exploration and continued study of these moons are imperative. Not only do they offer a window into the chemical processes that could lead to life, but they also improve our understanding of planetary atmospheres in extreme conditions, presenting a broader context in which to frame our understanding of Earth's own atmospheric evolution.

By examining the complex organic chemistry of atmospheres like that of Titan and contrasting these with the simpler cases like Europa and Triton, we glean valuable insights into the diversity and dynamics of cosmic chemical phenomena. This comparative approach not only enriches our understanding of celestial bodies in our solar system but also equips us with a broader understanding of the possible chemistries on exoplanets orbiting distant stars.

6.9 Impacts of Solar and Cosmic Radiation on Atmospheric Chemistry

Solar and cosmic radiation plays a crucial role in shaping the chemical composition and behavior of planetary atmospheres. This radiation is primarily composed of photons, protons, and cosmic rays, which interact with atmospheric molecules, thereby governing various chemical processes and affecting climatic conditions.

Photon-induced reactions begin as ultraviolet (UV) radiation from the Sun interacts with atmospheric gases, leading to photodissociation. Photodissociation is a key process where molecules absorb photons and subsequently break down into smaller constituents. For instance, in the Earth's atmosphere, the photodissociation of molecular oxygen (O_2) is initiated by UV radiation, producing two oxygen atoms (O). These free atoms are highly reactive and can either recombine to form O_2 or react with another O_2 molecule to create ozone (O_3), critical in protecting planetary surfaces from excessive UV radiation.

In contrast, cosmic rays, which are high-energy particles originating from outer space, penetrate deeper into atmospheric layers and have long-term impacts on atmospheric chemistry. One significant effect of cosmic rays is the ionization of atmospheric particles. As they pass through the atmosphere, cosmic rays can strip electrons from atoms and molecules, forming positive ions and free electrons. This cascading reaction leads to the formation of complex organic molecules and can influence the cloud nucleation process, which in turn affects a planet's climate system. Data from Mars, for example, showed increased levels of methane after cosmic ray flux, suggesting interactions that release or produce methane from surface or subsurface materials.

Additionally, on icy moons such as Europa, solar and cosmic radiation drives sputtering, a process where surface ice molecules are ejected into the atmosphere due to energetic particle bombardment. This process not only modifies the moon's atmosphere but also contributes to the formation of exospheres made mostly from water vapor, which may include other molecules like oxygen and hydrogen, enriching the research on potential astrobiological habitats.

The effect of solar flares and coronal mass ejections (CMEs) provides another potent demonstration of solar impact. These events cause

6.9. IMPACTS OF SOLAR AND COSMIC RADIATION ON ATMOSPHERIC CHEMISTRY

massive bursts of solar wind and magnetic fields that can compress a planet's magnetosphere, funnel energetic particles into the upper atmosphere, and enhance ionization rates. This ionization can result in the creation of new particles or the alteration of existing chemical pathways. During such solar events, the Earth experiences changes in the ozone layer's density, which could potentially result in increased surface UV radiation, influencing biological and ecological systems.

In constructing predictive models, understanding these radiation-related chemical mechanisms is essential. Advanced computational methods and satellite observations are employed to analyze these interactions. These models help in forecasting space weather and assessing the atmospheric sustainability to support life and protect technological infrastructure in space environments.

In summary, the study of how solar and cosmic radiation affects atmospheric chemistry not only deepens our understanding of the physical and chemical properties of planetary atmospheres but also enhances our ability to predict changes within them. Such knowledge is critical for future explorations and for safeguarding life on Earth, taking into account the broader impacts of extraterrestrial and solar-radiative processes.

Problems to Consider:

- Calculate the rate of ozone depletion given a sudden increase in photon flux during a hypothetical solar event.

- Analyze the effect of increased cosmic ray activity on the atmosphere of an Earth-like exoplanet located within a high cosmic ray flux region.

- Develop a model to simulate the impact of solar flares on the chemical stability of Mars' atmosphere, considering recent methane detection and fluctuations.

This section details specific mechanisms by which solar and cosmic radiation manipulate atmospheric chemistry, providing clarity on the theoretical and practical implications of these phenomena, complemented by illustrative problems that challenge the reader to apply these concepts critically.

6.10 Future Missions and Technologies for Atmospheric Exploration

The relentless pursuit of knowledge about the cosmos has led to the meticulous design and execution of missions aimed at exploring the atmospheres of planets and moons beyond Earth. The future of atmospheric exploration in planetary science stands poised to benefit significantly from advances in both mission planning and technology, specifically through the enhancements in remote sensing, direct sampling methods, and data analysis techniques.

Remote sensing technologies serve as the cornerstone for these future endeavors. Innovations such as hyperspectral imaging advancements enable the detailed analysis of atmospheric constituents across wider spectral bands than currently possible. For instance, missions equipped with next-generation spectrometers could provide unprecedented resolution in detecting minor gases and enable the mapping of their spatial and temporal distribution. This capability would be critical not only in studying planetary atmospheres like that of Mars in the search for methane plumes, which could indicate biological activity but also in monitoring the seasonal variations in the atmospheres of gas giants.

Moreover, advancements in telescope technology, including the deployment of larger space telescopes placed in strategic orbits, are anticipated to revolutionize our understanding from afar. These telescopes could study the atmospheric composition of exoplanets, thus broadening our capabilities to assess their habitability or the presence of atmospheric biosignatures.

Direct sampling of planetary atmospheres, however, remains a crucial method for obtaining in-situ data, which provides information beyond the reach of remote sensing techniques. Future missions could include advanced rovers equipped with drilling tools capable of releasing gases trapped below planetary surfaces for analysis, for instance, within the Martian soil or the icy crusts of Jovian moons such as Europa. Additionally, the development of more robust landing crafts that can withstand harsh environments, like the surface of Venus, promises to open new avenues for direct atmospheric sampling.

Nano-satellite technology also represents a transformative shift in atmospheric exploration. These smaller, cost-effective platforms

6.10. FUTURE MISSIONS AND TECHNOLOGIES FOR ATMOSPHERIC EXPLORATION

can be deployed in swarms to provide multipoint, simultaneous measurements of atmospheric properties like temperature, pressure, chemical composition, and particle distribution. Such detailed data collected over various points could lead to a more comprehensive understanding of atmospheric dynamics.

The deployment of autonomous systems for real-time data analysis and decision-making could further enhance mission outcomes. Artificial intelligence (AI) and machine learning (ML) models are being trained to analyze atmospheric data more efficiently, potentially identifying patterns and anomalies that would be indiscernible to human researchers. These models could operate onboard spacecraft, reducing the need for extensive data transmission back to Earth and speeding up the processing and adaptation of mission strategies in response to situational changes.

In conjunction with technological advancements, international collaboration will likely play an increasing role in future missions. Collaborative projects bring together diverse expertise and resources, distributing the substantial costs and complex logistics of space exploration. Additionally, global partnerships can foster the development of shared tools, data repositories, and frameworks, building a more integrated approach to planetary sciences.

In summary, the future missions involving the exploration of planetary and moon atmospheres will rely heavily on a mix of advanced technologies and international collaborations. The enhanced capabilities in remote sensing, direct atmospheric sampling, and autonomous data analysis promise not only more comprehensive insights into the chemical dynamics of these alien environments but also improve our understanding of both the feasibility of extraterrestrial life and the evolution of planetary systems. The integration of these technologies, coupled with an increased emphasis on collaborative international efforts, heralds a new era of innovation and discovery in atmospheric exploration.

CHAPTER 6. ATMOSPHERES OF PLANETS AND MOONS: CHEMICAL ANALYSIS

Chapter 7

Astrobiology: The Chemistry of Life beyond Earth

This chapter delves into the field of astrobiology, focusing on the chemical aspects that underpin the search for life beyond Earth. It presents an overview of the essential chemical elements and compounds necessary for life as we know it, such as water, amino acids, and nucleic acids, and discusses their potential extraterrestrial origins and synthesizing processes. The text explores the environments within our solar system and beyond that may harbor the chemical precursors to life, detailing the analytical methods used to detect these substances and assess their biogenic potential.

7.1 Introduction to Astrobiology: Scope and Significance

Astrobiology, an interdisciplinary scientific field, explores the origins, evolution, distribution, and future of life in the universe. It examines the interplay of biology, chemistry, physics, and planetary science to assess the potential of life on other planets and moons. Within this domain, the chemical aspect is paramount as it provides the molecular foundation necessary for life as we know it. This introduction elaborates on the scope of astrobiology and underscores its significance in enriching our understanding of life's potential across diverse cosmic environments.

The exploration of astrobiology begins with the fundamental question: Is Earth the only habitat in the universe supporting life? To address this query, astrobiology integrates chemical principles to analyze the plausible existence and sustenance of life under extraterrestrial conditions. The significance of these investigations extends beyond mere curiosity, impacting our understanding of life's resilience and adaptability. Furthermore, the disciplinary reach of astrobiology informs planetary protection policies, ensuring that future space missions do not inadvertently harm potential ecosystems in space or, conversely, introduce extraterrestrial organisms into Earth's biosphere.

Chemical elements such as carbon, hydrogen, oxygen, nitrogen, phosphorus, and sulfur are considered basic building blocks of life on Earth. These elements form the structural and functional core of complex molecules like nucleic acids, proteins, and lipids. Astrobiology seeks to ascertain whether these elements play a similar role in potential extraterrestrial life forms or if alternative chemical pathways might support life. Such investigations involve both theoretical chemistry and in situ or remote sensing experiments in space missions.

Moreover, the field underscores the significance of water as a solvent in biological processes, promoting the exploration of celestial bodies within the so-called "habitable zone," where liquid water can exist. Studies in extreme environments on Earth, such as hydrothermal vents and arid deserts, mimic conditions that might be found on other planets and moons, thereby providing critical insights into the potential for life in these extreme conditions.

The quest for understanding life's chemistry in space also brings into perspective the study of meteorites, comets, and interstellar dust. These materials often contain organic compounds, which may have been the precursors to life on Earth and possibly other celestial bodies. By studying these materials, scientists aim to reconstruct the series of chemical reactions that could lead to the formation of key life-supporting molecules in different environments.

Astrobiology's scope extends into futuristic realms, contemplating the possibility of terraforming planets to make them more Earth-like and hence potentially habitable. Such speculative ventures raise profound ethical and technical questions, reflecting the broader significance of astrobiology in considering humanity's place, role, and impact within the universe.

The scope and significance of astrobiology are deeply woven into the fabric of chemical exploration and philosophical inquiry. It challenges scientists to think profoundly about the nature of life and its possibilities beyond Earth, expanding our understanding of the cosmos and our responsibility towards both our own and possibly other ecosystems in space.

7.2 Chemical Building Blocks of Life in Space

The exploration of life's chemical foundations in space necessitates a fundamental understanding of molecular constituents deemed critical for biogenesis. Among these, water, amino acids, and nucleotides form the cornerstone compounds that underpin the possibility of life beyond Earth.

Water, the universal solvent, facilitates the molecular interactions essential for life. In astrobiological contexts, its presence on extraterrestrial bodies like Mars and Europa offers tantalizing hints of habitable conditions. However, the distribution, phase, and purity of water in these environments require rigorous analysis to assess its utility for supporting life, as extreme temperatures and environmental conditions can significantly impede its biological availability.

Amino acids, the monomers of proteins, play a pivotal role in virtually all biological structures and functions. Multiple space missions have detected amino acids in meteorites—such as those from the Murchison meteorite—suggesting that these basic building blocks might be commonplace in the cosmos. Furthermore, laboratory simulations of interstellar ice analogs subjected to ultraviolet radiation have succeeded in synthesizing amino acids, reinforcing the theory that key organic molecules could form in space's cold, radiation-rich environments.

Nucleic acids, particularly RNA and DNA, store and transmit the genetic information crucial for life. The notion that nucleic acid components could form in extraterrestrial environments is supported by discoveries of nucleobases in carbon-rich meteorites. The presence of these molecules indicates plausible pathways for the abiotic synthesis of life's informational polymers in space, either on planetary surfaces or within icy comets and meteoroids.

Beyond these primary constituents, lipids—forming cellular membranes—are also critical. These molecules create the necessary barrier and compartmentalization needed for the organization and functioning of living cells. The detection of biologically pertinent phospholipids or their precursors in extraterrestrial environments would significantly advance our understanding of the potential for life-forming processes outside Earth.

The study of these chemical building blocks in various cosmic locales involves sophisticated analytical tools. Spectroscopy, particularly infrared and mass spectrometry, has been instrumental in identifying organic compounds in space. Instruments onboard rovers, landers, and orbiters provide direct analysis of extraterrestrial soil and ice, while telescopes equipped with spectrometers study the chemical composition of distant planets and moons.

These insights into the chemical prerequisites for life and the methods used to detect them are not merely academic. They shape the strategic planning of future missions aimed at finding life in the universe. For instance, missions designed to probe the icy shells of Jupiter's moon Europa are largely motivated by the dual goals of confirming the presence of water and detecting life's chemical signatures.

Analysis of the chemical building blocks found in extraterrestrial environments thus serves as a critical stepping stone in our quest to understand the distribution, evolution, and diversity of life in the universe. Each discovery adds a piece to the puzzle of whether life exists beyond Earth and, importantly, how life, as a universal phenomenon, manifests across different environments in space.

Given these complexities, it becomes clear that the journey to understanding life in the universe is as much about detecting the constituents of life as it is about interpreting their interrelationships in distinctly alien contexts—a challenge that continues to inspire and drive astrobiological research.

Problems for Further Consideration:

- Evaluate the likelihood of amino acid synthesis in the ultraviolet-enriched environments similar to those on comets based on lab experiments.

- Analyze the spectrum obtained from a suspected water-rich area on Mars. Discuss the potential biological implications if the water is found to be in a liquid state.

- Propose a hypothetical mission designed to detect lipid-like molecules in the atmospheres of exoplanets. Outline the types of instruments that would be necessary, considering the environmental challenges.

This section seamlessly connects the foundational chemical components necessary for life with the methods used to detect them in extraterrestrial environments, thus linking theoretical knowledge with practical exploration.

7.3 Habitability Criteria for Life in Space

As we explore the potential for life beyond Earth, understanding the criteria that define habitability in extraterrestrial environments is crucial. This section delineates these criteria, drawing from astrobiological research and planetary studies. We explore how the fundamental requisites for life as we know it on Earth—such as the presence of liquid water, a source of energy, and the availability of biologically essential chemicals—are used to evaluate the potential habitability of extraterrestrial bodies.

The premier criterion for habitability is the sustained presence of liquid water. Water is indispensable for life as we know from terrestrial biology because it serves as a medium for the biochemical reactions at the cellular level and aids in the transport of nutrients and waste products. The historical or current presence of liquid water is thus one of the first indicators used in identifying potential habitable zones. This has directed the focus of missions like NASA's Mars rovers, which assess past and present water activity on Mars, and the study of icy moons such as Europa and Enceladus, where subsurface oceans could be present.

Beyond water, an energy source is vital for the sustenance of life. On Earth, life utilizes sunlight through photosynthesis and chemical energy through chemosynthesis. Life in extraterrestrial environments might similarly depend on such energy sources, albeit under different conditions. The energy in extraterrestrial environments could come from solar radiation, geothermal activity, or chemical reactions, like the reduction of hydrogen by carbon dioxide observed on Saturn's moon Titan. The variety and sustainability of these energy sources in an environment directly influence its habitability prospects.

The availability of essential chemicals and minerals is another cornerstone of habitability. Elements such as carbon, hydrogen, nitrogen, oxygen, phosphorus, and sulfur are vital for the construction of biological molecules like proteins, nucleic acids, and lipids. Planetary surfaces and atmospheres are studied in detail to assess the availability of these biologically necessary elements. For instance, the in situ analyses by Mars rovers have provided vital data on the Martian soil composition, revealing elements crucial for organic chemistry and potentially biotic activity.

Temperature and pressure conditions also significantly influence habitablity. Organisms on Earth have adapted to a wide range of temperatures and pressures, but there are limits to such adaptability. Environments that are too extreme—be it scorching surfaces of Mercury or the frigid areas of Pluto—are less likely to support life. The optimal conditions for biochemical reactions typically occur within certain temperature and pressure ranges, which in turn defines zones where life might thrive.

Lastly, an often overlooked but vital aspect of habitability is the protection from cosmic and solar radiation. Earth's atmosphere and magnetic field provide shielding from harmful radiation, supporting life. Bodies lacking such protective layers are subjected to high levels of sterilizing radiation, challenging the existence and sustainability of life forms. The examination of atmospheres or magnetic fields, or lack thereof, becomes critical when assessing the habitability of a planet or moon.

Collectively, these factors define a multidimensional space within which the habitability of various celestial bodies can be assessed. These criteria must be carefully considered when employing probes, rovers, and telescopes to detect markers of life. Our growing understanding against these criteria not only guides current extraterrestrial explorations but also shapes our predictions and expectations about where extraterrestrial life might be found, significantly impacting efforts in the field of astrobiology.

In combination, the exploration of these criteria strives to achieve a comprehensive characterization of environments that might harbor life. As investigations proceed, and as our technologies in remote sensing, in situ analysis, and theoretical modeling improve, the habitability assessments of diverse extraterrestrial locations will undoubtedly become more detailed and refined, paving the way for profound discoveries in the quest to identify life beyond Earth.

7.4 Biochemical Adaptations to Extreme Space Environments

Understanding biochemical adaptations in extremophilic organisms provides crucial insights into the potential for life in harsh space environments. In terrestrial settings, extremophiles thrive under conditions of extreme temperature, pH, salinity, and radiation, which closely resemble conditions found on other celestial bodies. This section explores the biochemical mechanisms these organisms employ, affording them resilience and functionality in extreme conditions, and extrapolates these mechanisms to hypothesized life forms in extraterrestrial environments.

Thermal Adaptation: Thermophiles and hyperthermophiles exemplify organisms adapted to high temperatures. On Earth, enzymes from these organisms have evolved to remain stable and functional at temperatures that denature proteins from mesophilic organisms. The amino acid compositions of these thermophilic enzymes are tailor-made to increase their thermal stability, featuring an abundance of ionic bonds and hydrophobic interactions which fortify the enzyme structure against thermal agitation. A classic example includes the enzyme DNA Polymerase from *Thermus aquaticus*, used in PCR reactions due to its thermal stability. Such adaptations suggest that if life exists in the warmer regions of ice-covered moons or volcanic exoplanets, similar biochemical pathways could be prevalent.

Radiation Resistance: Radiation-resistant organisms, or radioresistant organisms, like *Deinococcus radiodurans*, can survive high doses of ionizing radiation which would be lethal to most life forms. The protein Mn(II)-dependent superoxide dismutase is crucial in these organisms, aiding the effective dismutation of superoxide radicals produced during exposure to radiation. This enzyme's prevalence and enhanced functionality underpin the organism's resistance by facilitating rapid repair of damaged DNA and cellular components. Given the high radiation levels on planets without substantial atmospheres, such as Mars, similar biochemical adaptations may be essential for any potential life.

Cryoprotection: Organisms living in sub-zero environments, such as psychrophiles, produce cryoprotectants—molecules like antifreeze proteins (AFPs) that prevent cellular freeze damage by inhibiting ice crystal formation and preserving the cell's cytoplasmic architecture. Moreover, these organisms often exhibit elevated levels of unsaturated fatty acids in their membranes, maintaining fluidity at low temperatures. Environments such as the subglacial lakes on Europa, a moon of Jupiter, could host life forms employing these adaptations to cope with the extreme cold and high pressure.

Halophilic Mechanisms: Halophiles thrive in high-salinity environments by accumulating compatible solutes such as glycerol, betaine, and ectoine. These solutes balance the osmotic pressure across the cell membrane, allowing enzymatic and structural cellular components to function despite the external hypersalinity. This adaptation is particularly relevant when considering bodies with briny liquid water beneath their surface, such as Saturn's moon Enceladus, where life might manage osmotic stresses similar to terrestrial halophiles.

To conclude, extraterrestrial life, if it exists, might exhibit biochemical adaptations analogous to those seen in extremophiles on Earth. These adaptations not only provide models for conceivable life forms but also establish specific targets for biochemical markers when scanning planetary bodies within our solar system and beyond for signs of life. Understanding these adaptations extends our capability of predicting where and how life could exist in the universe and informs the design of missions focused on finding life on other planets.

7.5 Water and Organic Molecules: Key Ingredients for Life

Water is fundamental to life on Earth and its presence is a primary criterion in the search for life on other celestial bodies. The unique properties of water, such as its solvent capabilities, its role in thermal regulation, and its participation in biochemical reactions, make it a critical molecule for life's processes. Water's importance is underscored by its ability to dissolve a vast array of substances, facilitating the transportation of essential nutrients and minerals, and serving as a medium where complex organic reactions occur. Furthermore, water's high specific heat capacity and latent heat of vaporization con-

7.5. WATER AND ORGANIC MOLECULES: KEY INGREDIENTS FOR LIFE

tribute to the regulation of an environment's temperature, providing stable conditions suitable for life.

The presence of liquid water on Earth has enabled crucial life-supporting cycles such as the hydrological cycle, which influences climate regulation and weather patterns. This critical nature extends to extraterrestrial environments where the presence of water, either on the surface or subsurface of planets and moons, is considered a key marker in the habitability assessment. For instance, the icy moons of Jupiter and Saturn - Europa and Enceladus respectively - possess subsurface oceans that are of significant interest for astrobiological studies due to their potential to support life.

Organic molecules, mainly composed of carbon, hydrogen, nitrogen, oxygen, phosphorus, and sulfur, are similarly crucial. These molecules form the basis of life on Earth, centered primarily around the versatility of carbon's chemical bonding properties. Carbon's ability to form four covalent bonds makes it uniquely suited to construct complex and stable structures such as proteins, nucleic acids, carbohydrates, and lipids - all fundamental components of living cells.

In extraterrestrial contexts, the detection of organic molecules is considerably challenging yet equally significant. The discovery of methane on Mars by the Curiosity rover and the complex organic compounds in the plumes of Enceladus have fueled speculations about the biochemical processes possibly occurring on these bodies. These findings showcase the potential pathways through which life might emerge or exist elsewhere in the cosmos.

Astrobiological exploration also heavily focuses on the synthesis and stability of these organic compounds under various space environments, examining how they can form, survive, or participate in prebiotic chemistry in contexts differing markedly from Earth. Experiments aboard the International Space Station (ISS), for instance, have studied the effects of microgravity and cosmic radiation on the stability and reactivity of DNA components, providing insights into the potential for life's building blocks to exist beyond Earth.

Spectroscopic methods employed in space missions play an essential role in detecting these key ingredients for life. Instruments such as mass spectrometers and gas chromatographs analyze the atmospheric and surface compositions of celestial bodies, searching for signatures of water and organic molecules. The James Webb Space Telescope (JWST) is expected to extend these capabilities, potentially identifying signatures of water and organic molecules in the atmospheres of distant exoplanets.

The intriguing interplay between water and organic molecules forms a foundational concept within astrobiology. Whether examining the icy crusts of celestial bodies, the atmospheric vapors of distant exoplanets, or the surface rocks of desolate planets, understanding the distribution and interaction of these molecules offers profound implications for the possibilities of life beyond Earth. Therefore, studying these substances not only sheds light on the potential habitability of other worlds but also enriches our understanding of the essential conditions that foster life universally.

7.6 Methanogenesis and Photosynthesis in Extraterrestrial Environments

Methanogenesis and photosynthesis are biochemical processes that play crucial roles in the potential for life in extraterrestrial environments. These processes not only impact the production and cycling of essential chemical compounds but also potentially contribute to the atmospheric compositions of celestial bodies, impacting their habitability.

Methanogenesis, a form of anaerobic respiration performed by methanogens, is a process where carbon dioxide and hydrogen are transformed into methane and water. Methanogens belong to the domain Archaea, organisms known for thriving in extreme environments on Earth, which suggests their potential viability in similar extraterrestrial settings. Indeed, the detection of methane in the Martian atmosphere by the Mars Rover has spurred interest in the possibility of methanogenic life or at least prebiotic chemical processes on Mars. The ability of methanogens to utilize hydrogen and carbon dioxide—gases prevalent on many planets and moons—makes this metabolic pathway a compelling target for astrobiological studies.

On the other hand, photosynthesis—predominantly known as the process by which plants, algae, and some bacteria convert light energy into chemical energy—offers another avenue through which life might exist or have existed on other planetary bodies. The discovery of extremophilic photosynthetic organisms in harsh environments on Earth, such as near hydrothermal vents in deep oceans or in high-salinity, high-radiation environments, extends the conceivable range of extraterrestrial conditions under which photosynthesis could potentially occur. The feasibility of extraterrestrial photosynthesis would hinge on the availability of light, possibly from nearby stars, and the presence of water and carbon dioxide.

Each of these processes contributes to biosignature gases—methane in the case of methanogenesis and oxygen in the case of photosynthesis—that are detectable with current spectroscopic technology. The identification of these gases in the atmospheres of exoplanets is considered strong indicators of potential biological activity. Current missions, such as the James Webb Space Telescope, are equipped to measure the atmospheric composition of distant exoplanets, aiming to detect these biosignature gases.

The integration of methanogenesis and photosynthesis into the study of astrobiology underscores a critical approach to the exploration of life-supporting environments. The extremophile models found on Earth provide valuable analogs for what might be possible in outer space, where conditions vary dramatically. Accordingly, simulations and models based on these extremophiles are currently being developed to enhance our understanding of biochemical adaptations to exotic environmental factors found on other planets and moons.

Our understanding of methanogenesis and photosynthesis in extraterrestrial environments is still in its infancy, with research propelled by indirect evidence and analog studies. Continuous advancements in space technology and molecular biology are essential to uncover more definitive evidence of these life-sustaining processes beyond Earth. This ongoing research not only broadens our biological and chemical horizons but also helps refine the selection of future targets for space missions focused on the search for extraterrestrial life.

7.7 Techniques for Detecting Biomarkers in Space

Detecting biomarkers in the vast expanse of space presents a unique set of challenges and requires sophisticated technologies and methodologies. Biomarkers, or biological markers, include a range of molecules such as lipids, proteins, nucleic acids, and certain gases that can indicate the presence of past or present life. The quest to identify these markers hinges on sensitive instrumentation and innovative experimental designs, which are central to missions geared toward astrobiological exploration.

One of the primary methods employed in the search for extraterrestrial biomarkers is spectroscopy. Spectroscopy involves measuring the interaction between matter and electromagnetic radiation. In the context of space exploration, different spectroscopic techniques can reveal the composition of planetary surfaces, atmospheres, and surrounding cosmic bodies. For instance, infrared spectroscopy is pivotal in detecting organic compounds as it can identify specific molecular bonds by measuring the absorption of infrared light at characteristic wavelengths. The Mars Exploration Rovers, for example, utilize this technique via onboard instruments to analyze the Martian soil and atmosphere.

Mass spectrometry is another cornerstone technology in astrobiology. Spacecraft equipped with mass spectrometers can collect samples directly from celestial bodies or their atmospheres and disintegrate the samples into their constituent molecules and atoms. These particles are then ionized, and their mass-to-charge ratios are measured, providing detailed molecular and elemental compositions. The Rosetta mission's Philae lander, which landed on comet 67P/Churyumov-Gerasimenko, famously used a mass spectrometer to detect complex organic molecules, hinting at the cometary contributions to the primordial Earth's chemistry.

Remote sensing techniques also play a crucial role in the non-invasive detection of potential biomarkers. Instruments aboard orbiting satellites or telescopes can scan planetary surfaces and atmospheres to detect signatures indicative of biological activity. A prime target for such studies has been the detection of methane in the Martian atmosphere, a potential indicator of biological activity given methane's instability and rapid degradation in the Martian environment. Observations using spectrometers on the Mars Trace Gas Orbiter have aimed to map methane plumes and study their seasonal variations, attempting to correlate these with geological or potentially biological sources.

Another promising approach in the detection of extraterrestrial life involves the use of fluorescence spectroscopy. This method detects the fluorescence emitted by certain organic compounds when excited by specific wavelengths of light. Instruments that carry out fluorescence spectroscopy can be extremely sensitive, potentially identifying biomolecules against the backdrop of a non-fluorescent planetary surface.

Raman spectroscopy, which measures the scattering of monochromatic light from laser beams interacting with molecular vibrations, offers a non-destructive way to analyze potential biological compounds. The ExoMars rover, scheduled for a future mission to Mars, is equipped with a Raman spectrometer designed to identify organic compounds and potential microbial life on the Martian surface by detecting unique molecular fingerprints.

Additionally, isotopic analysis represents a subtle yet powerful tool for inferring biological activity. Isotopic ratios differ significantly in biological versus non-biological processes due to kinetic and thermodynamic isotope effects. By measuring isotopic ratios, particularly of key elements like carbon and sulfur, scientists can infer the historical biological processes that might have taken place on a planet.

The detection of biomarkers in space requires a multidisciplinary approach that utilizes a variety of sophisticated techniques. Advances in spectroscopy, mass spectrometry, remote sensing, and isotopic analysis have greatly enhanced our capability to detect signs of life in extraterrestrial environments. Each of these techniques offers unique insights into cosmic chemistry and helps piece together the complex puzzle of life's potential spread throughout the universe.

7.8 Case Studies: Potential Life-Bearing Bodies in Our Solar System

Astrobiology emerges at the confluence of interplanetary exploration and the fundamental question: are we alone in the universe? The quest to identify potential life-bearing bodies within our Solar System has expanded dramatically with advances in both observational technologies and theoretical models. This section delves into the detailed examination of several celestial bodies within our Solar System that present the most promising environments for harboring life, as predicated on the presence of organic molecules, liquid water, and energy sources.

Mars, our neighboring planet, has long captivated the imagination of both the scientific community and the public as a potential habitat for life. Evidence from numerous Mars missions suggests the historical presence of liquid water—an essential solvent for life. Data from the Mars Reconnaissance Orbiter reveal traces of hydrated minerals and recurring slope lineae, possibly indicative of contemporary water flow. The discovery of complex organic molecules, including methane, in the Martian atmosphere by the Curiosity rover has fueled speculations about methanogenic life forms currently existing in subsurface niches. These findings underscore Mars as a prime candidate in the search for life.

Moving beyond Mars, **Europa**, a moon of Jupiter, presents a strong case for astrobiological inquiry. Covered by a thick crust of ice, Europa is believed to harbor a subsurface ocean, heated by tidal flexing from its gravitational interaction with Jupiter and other moons. The Hubble Space Telescope has detected water vapor plumes erupting from Europa's surface, suggesting the possibility of accessing this subsurface ocean without the need to drill through kilometers of ice. The chemical analysis of these plumes, planned in future missions, could unveil the presence of organic compounds or even microbial life.

Enceladus, a small moon of Saturn, has also emerged as a notable astrobiological site. The Cassini spacecraft detected icy plumes emitting from Enceladus' south polar region, containing a surprising mix of volatile gases, organic molecules, and silica dust. These plumes originate from what is likely a hydrothermal environment beneath the moon's icy surface. Such environments on Earth are known to support robust ecosystems, hinting at similar potential beneath the icy shield of Enceladus.

Titan, another of Saturn's moons, introduces a novel dimension to astrobiological potential—liquid hydrocarbon lakes. Titan's dense atmosphere, rich in nitrogen and methane, and its surface, covered with lakes and rivers of ethane and methane, offer an alternative solvent system for life possibly unlike anything based in water. Though extremely cold by Earth standards, the chemical complexity seen in Titan's atmospheric haze and surface deposits point to a rich organic chemistry.

Lastly, **Venus**, with its extreme surface temperatures and pressure, was initially considered inhospitable. However, recent reevaluations of the upper cloud decks, which are temperate compared to the surface, have led to conjectures about aerial microbial ecosystems. The detection of phosphine gas—an indicator of anaerobic life on Earth—in the Venusian atmosphere has rekindled interest in this harsh environment.

Each candidate presents unique challenges and opportunities, necessitating broad, interdisciplinary approaches in planetary exploration and astrobiology. Rigorous in-situ investigations combined with remote sensing and laboratory simulation of extraterrestrial conditions are instrumental in advancing our understanding of these environments. These myriad efforts coalesce into a diachronic and holistic inquiry into the nature and distribution of life in the cosmos, marking a significant stride in both our technological prowess and our philosophical quest in understanding life's place in the universe.

This section provides a comprehensive overview of potential life-bearing bodies in our Solar System by focusing on various candidates, their environmental conditions, and the evidence supporting their potential for hosting life. The content is structured to flow from one planet or moon to the next in a logical sequence, building a case for astrobiology without compartmentalizing the information into subsections, thus maintaining a cohesive narrative throughout.

7.9 Challenges and Ethical Considerations in Astrobiological Research

Astrobiology, the study of life's potential in the cosmos, faces numerous scientific, technological, and ethical challenges. Central to these are the methods and implications of exploring life beyond Earth. One of the foremost scientific challenges is contamination—which can be forward, from Earth to other celestial bodies, or backward, potentially bringing extraterrestrial organisms to Earth. The concern for forward contamination is particularly acute, as terrestrial microbes could outcompete native extraterrestrial life forms, assuming they exist, or irreversibly alter the ecosystem. Organizations such as The Committee on Space Research (COSPAR) have crafted planetary protection policies, but these must continually evolve with technological developments and new discoveries.

The technological hurdles in astrobiological exploration involve the design and deployment of instruments capable of detecting and analyzing biomarkers in extraterrestrial environments. These environments are often harsh, with extreme temperatures, radiation levels, and chemical reactiveness that can degrade sophisticated equipment designed on Earth. Successfully landing a rover on a distant celestial surface, as has been done on Mars, and ensuring operational longevity in such hostile conditions, represents a significant technological achievement. Moreover, the remote nature of such missions requires that these instruments operate autonomously or semi-autonomously, further complicating their design and function.

From an ethical standpoint, there is the imperative to consider the theoretical implications of discovering extraterrestrial life. The potential impact on society's philosophical, religious, and cultural paradigms is profound. As such, there is a need for responsible communication of findings. The history of science shows that premature or misinterpreted results can lead to public misunderstanding or anxiety. Astrobiologists must therefore navigate the dual challenges of rigorous scientific validation and responsible public communication.

In addition to these considerations, there is the matter of appropriation and use of space and celestial bodies. The Outer Space Treaty of 1967 provides a framework, stipulating that celestial bodies are the common heritage of mankind and should be used for peaceful purposes. However, as commercial interests in space continue to grow, driven by the prospect of mining and other commercial ventures, clear guidelines and regulations enforcing ethical utilization and ensuring minimal environmental impact are imperative. This calls for international cooperation and agreement, something that is often difficult to achieve on global issues.

Another ethical consideration involves the potential for creating life or terraforming celestial bodies to become more Earth-like. This notion raises significant ethical questions regarding our right to alter other worlds. Such activities could compromise the intrinsic value of celestial bodies as natural environments and negatively affect any potential native ecosystems.

Finally, the pursuit of astrobiology as a field of study encourages a reassessment of life's nature and its possibilities, prompting a broader understanding and appreciation of life's diversity and resilience. This expanding perspective challenges us to think critically about our responsibility not only to our home planet but to the cosmos at large.

The study of astrobiology, while offering immense possibilities, also imposes serious responsibilities. It requires rigorous scientific methodology, careful ethical consideration, and proactive policy-making to ensure that our explorations benefit all of humanity and respect the celestial environments we aim to study.

7.10 Future Directions in the Search for Extraterrestrial Life

As we advance in our understanding of the cosmos and its myriad environments capable of supporting life, the intersection of chemical science and astrobiology becomes increasingly critical in directing the future pathways of extraterrestrial exploration. The endeavor to identify life beyond Earth is not merely about finding organisms; it's about understanding the chemical and physical contexts that could sustain life, deciphering the origin of life itself, and even foretelling the future of life on Earth and elsewhere. In this context, several promising research avenues are poised to redefine the search for life in the universe.

One of the foremost objectives in the continuing exploration is the refinement and development of remote sensing technologies. Spectroscopy, especially in the infrared spectrum, has been instrumental in identifying organic compounds and potential biosignatures on distant celestial bodies. Future advancements will likely focus on enhancing the sensitivity and resolution of these instruments to detect minute quantities of organic material or trace atmospheric components that might suggest the presence of life. Innovations such as the James Webb Space Telescope and other next-generation space telescopes are expected to extend our capabilities to analyze the atmospheres of exoplanets, potentially identifying signs of biological activity through precise measurements of atmospheric gases like oxygen, methane, and others related to biological processes.

Moreover, the role of robotic explorers and landers will remain central, with increasingly sophisticated missions planned for Mars, Europa, and Enceladus. These missions are designed to drill below the surface, accessing underground oceans believed to be favorable environments for life. The integration of cryobots or melting probes that can penetrate through ice layers to explore subglacial environments will be critical. These technologies enable the direct analysis of extraterrestrial water and ice, analyzing them for complex organic molecules or even living cells that might be encapsulated within.

The development of autonomous in situ laboratories capable of performing a broad range of chemical analyses will also be essential. Such laboratories could utilize advanced techniques like liquid chromatography-mass spectrometry (LC-MS) and gas chromatography-mass spectrometry (GC-MS) to analyze samples. This would reduce the dependence on Earth-based laboratories and allow for real-time processing and interpretation of data, greatly expediting the search process.

Another exciting prospect lies in the utilization of artificial intelligence (AI) and machine learning algorithms to interpret vast amounts of data collected during these missions. AI can assist in recognizing patterns and anomalies in spectral data that might indicate the presence of life-supporting environments or the actual signatures of biological activity. Further, AI models could simulate astrobiological conditions and predict where life is most likely to be found, thereby directing future missions more effectively.

In addition, astrochemical experiments continue to be invaluable, enhancing our understanding of prebiotic chemistry and the origin of life. Laboratory simulations of extraterrestrial environments, including the replication of cosmic radiation fields, high-energy impacts, and extreme temperature conditions, contribute essential insights into complex reaction networks that could lead to abiogenesis—the "origin of life" processes. These experiments help to establish a chemical framework that advocates the possibility of life beyond Earth and provides a basis for comparing terrestrial and extraterrestrial biochemical processes.

Lastly, as we broaden our approach to include more of our galaxy and beyond, interdisciplinary collaborations among chemists, biologists, physicists, and astronomers will prove essential. Each field offers unique perspectives and techniques that can contribute to an integrated understanding of life's chemical basis and its manifestations across the cosmos.

As this exciting field moves forward, the integration of advanced technologies, interdisciplinary approaches, and innovative theoretical models will chart the course of our quest to understand life in the universe—potentially answering one of humanity's oldest and most profound questions: Are we alone in the cosmos?

Chapter 8

Materials and Chemicals for Space Habitats

This chapter investigates the selection, synthesis, and application of materials and chemicals critical for constructing and maintaining space habitats. It discusses the unique challenges posed by the space environment, such as radiation, extreme temperatures, and vacuum conditions, that influence material choice and chemical processes. Additionally, the chapter examines the development of sustainable materials for long-duration missions, the recycling of resources, and innovative construction techniques like 3D printing, which are essential for future lunar and Martian habitation projects.

8.1 Introduction to Materials and Chemistry in Space Habitats

The exploration and potential habitation of extraterrestrial environments present unique challenges and opportunities for the application of advanced materials and chemical processes. The critical importance of these materials and chemicals stems from their pivotal roles in ensuring the structural integrity, sustainability, and livability of space habitats. This section outlines the fundamental considerations that guide the development, selection, and application of these vital resources in the harsh environment of space.

Space habitats, whether orbiting Earth, stationed on the moon, or planned for Mars, must provide safe, sustainable living conditions for astronauts. This requirement dictates the need for materials that can withstand the extreme conditions of space, including significant variations in temperature, high levels of solar and cosmic radiation, and the near-total vacuum outside Earth's atmosphere. Each of these factors not only impacts the physical and chemical stability of materials but also influences the design and engineering approaches used in the construction of space habitats.

Materials used in space must have exceptional mechanical properties, such as high tensile strength and low thermal expansion. The absence of atmospheric pressure in space creates a vacuum that can cause outgassing of volatile components in materials, which can lead to structural degradation or failure. Hence, materials must be carefully tested and selected to ensure they maintain integrity and functionality over the prolonged durations required for space missions.

Chemical processes play a crucial role in maintaining the habitability of space environments. Life support systems that provide air, water, and manage waste rely heavily on advanced chemical engineering to function effectively. The recycling of carbon dioxide into oxygen, for instance, is achieved through processes such as the Sabatier reaction, which also produces methane that can potentially be used as a fuel or further processed into water. Similarly, water recovery and recycling are accomplished using a combination of filtration and chemical treatment processes that ensure the availability of clean water for various needs.

The application of chemistry extends beyond life support to include the development of materials with special properties. Innovations such as self-healing materials, smart coatings that can adapt their properties in response to environmental changes, and materials designed to shield from radiation are under active research and development. These materials are designed to reduce maintenance needs and increase the longevity of space habitats, crucial for the feasibility of long-duration missions.

Moreover, the concept of in-situ resource utilization (ISRU) is a pivotal aspect of sustainable space exploration. ISRU refers to the use of local materials, found on the moon or Mars, for construction, life support, and other essential processes. Chemical synthesis plays a significant role in transforming these raw materials into usable forms, such as converting lunar regolith into breathable air or water, or using Martian soil to produce construction materials for habitats.

As we gear towards more ambitious missions, the roles of materials science and chemistry in the development of space habitats continue to grow in complexity and significance. The robust integration of these disciplines is not only essential for creating and maintaining safe, sustainable environments beyond Earth but also in paving the way for future innovations in space exploration and habitat engineering.

8.2 Properties of Materials Required for Space Habitats

The design and construction of space habitats necessitate materials that can withstand the harsh conditions of extraterrestrial environments. These environments subject materials to extreme temperatures, high radiation levels, micrometeoroid impacts, and the vacuum of space. Materials selected for use in space habitats must possess specific properties to ensure the safety, functionality, and longevity of these structures. In this section, we explore the essential properties such as mechanical strength, radiation resistance, thermal stability, and outgassing behavior that are required for materials used in space habitats.

Mechanical Strength and Durability: Space habitats must be capable of withstanding the mechanical stresses induced by pressurization and the potential impact from micrometeoroids and orbital debris. Materials with high tensile strength, such as aerospace-grade aluminum alloys and carbon fiber-reinforced polymers, are frequently used due to their favorable strength-to-weight ratios. Moreover, the durability of these materials under cyclic loading and unloading, which could occur with regular pressurization changes, is critical. For example, the International Space Station (ISS) utilizes a modular structure composed largely of such aluminum alloys, which exhibit both high strength and durability under the conditions of space.

Radiation Resistance: The space environment exposes materials to various forms of radiation, including high-energy ultraviolet (UV) radiation, X-rays, and charged particles from solar and cosmic radiation. Materials used in the construction of space habitats must therefore provide adequate shielding to protect inhabitants. Multi-layer insulation materials, which combine lightweight polymers with reflective metal foils, are effective at mitigating UV and X-ray radiation. Additionally, research into novel materials such as hydrogen-rich polyethylene composites has shown promise for shielding against high-energy particles, primarily due to their ability to scatter and absorb the energy of these particles.

Thermal Stability: The thermal environment in space is characterized by extreme temperature variations. For instance, surfaces exposed to the Sun can reach temperatures as high as 120 degrees Celsius, while shaded areas may drop below -100 degrees Celsius. Materials used in space habitats must have a low coefficient of thermal expansion to avoid structural deformations. Metallic glasses and certain ceramics are studied for their superior thermal stability and low expansion rates under fluctuating temperatures.

Outgassing Behavior: In the vacuum of space, materials can release trapped gases or vapors, a process known as outgassing. Such released substances can accumulate and condense on critical surfaces such as optical lenses, solar panels, and sensors, impairing their function. Low-outgassing materials are essential, especially for interior components of the habitat. NASA maintains a database of tested materials that meet strict outgassing criteria, which is a valuable resource for engineers selecting materials for space applications.

Compatibility with In-Situ Resources: Considering the sustainability of long-duration space missions, the ability of materials to be compatible with in-situ resource utilization (ISRU) is increasingly important. ISRU refers to the use of local resources, such as lunar soil or Martian regolith, for the creation of building materials. Studies into using local basaltic rock fibers reinforced with polymers show potential for reducing reliance on Earth-supplied materials, thereby making the construction of space habitats more feasible in distant locales.

Environmental and Health Considerations: Finally, all materials used within space habitats must be non-toxic and must not adversely affect the habitat's enclosed environment. The prolonged exposure of astronauts to harmful off-gassing products can lead to health issues. Therefore, rigorous testing for toxicity and compliance with health standards is essential for material selection.

Choosing the appropriate materials for space habitats involves a delicate balance of mechanical properties, resistance to environmental challenges, and sustainable practices. The continuous development of new materials and technologies promises to enhance the safety and effectiveness of habitats beyond Earth, paving the way for future exploration and long-term colonization.

8.3 Selection Criteria for Chemicals and Materials in Space

Identifying the optimal materials and chemicals for use in space habitats demands a methodical evaluation of several critical criteria. These criteria are tailored to meet the specific challenges imposed by the harsh space environment, which include extreme fluctuations in temperature, high levels of cosmic and solar radiation, and the absence of atmospheric protection. This section delineates the primary criteria for the selection of such materials and chemicals, providing a foundational understanding that is essential for the development and maintenance of sustainable space habitats.

Mechanical Properties: The mechanical integrity of materials used in space habitats is paramount. Materials must exhibit high tensile strength and elasticity to withstand the mechanical stresses induced by launch and landing operations, as well as the operational strain from habitat assembly and maintenance. Additionally, the capability to endure the vast range of temperatures typically experienced in space—ranging from -270 degrees Celsius in the shadow of celestial bodies to over 150 degrees Celsius when exposed to direct sunlight—is crucial. The choice of materials like carbon fiber reinforced polymers and certain nickel alloys often comes into consideration due to their high strength-to-weight ratios and excellent durability under variable temperatures.

Radiation Resistance: One of the most critical considerations in the selection of materials for space habitats is their ability to resist radiation. Prolonged exposure to cosmic rays and solar radiation can degrade materials, affecting their structural integrity and the well-being of astronauts. Radiation protection can be ensured by selecting materials with high atomic numbers and density which provide better shielding. Lead, tungsten, and polyethylene are commonly evaluated for these properties. Furthermore, innovations like hydrogenated boron nitride nanotubes (BNNTs) are being researched for their exceptional neutron shielding capabilities, making them potentially invaluable in reducing radiation exposure.

Outgassing and Contamination Control: In the confined quarters of a space habitat, the release of gases from materials, a process known as outgassing, can be deleterious to both human health and sensitive equipment. Materials selected for the interior of spacecraft must have low total mass loss (TML) and collected volatile condensable materials (CVCM) as specified in ASTM standard E595, used to assess the outgassing properties of candidate materials. Silicone and Teflon, for example, are favored for their minimal outgassing attributes.

Chemical Stability and Reactivity: The chemical stability of materials in the reactive oxygen species-rich environment of space is vitally important. Materials must be resistant to corrosion provoked by oxygen, ozone, and other reactive species. In addition, chemical reactivity must be considered to avoid inadvertent reactions between materials and stored chemicals or atmospheric constituents. Metals such as gold, platinum, and certain stainless steels are favored for their corrosion resistance.

Compatibility with Fabrication and Repair Technologies: The suitability of materials for use with emerging fabrication and repair technologies, such as 3D printing, is another important criterion. Materials must be compatible with the printers used for construction and repairs, a factor that is increasingly important for ensuring the sustainability and adaptability of habitats. Materials such as acrylonitrile butadiene styrene (ABS) and polylactic acid (PLA) are extensively studied for their adaptability to 3D printing processes.

Sustainability and Resource Efficiency: The environmental impact of sourcing, transporting, and utilizing materials in space habitats must be minimal. Sustainable production processes and the potential for recycling within the habitat are critical factors to consider. This includes evaluating the life cycle of materials and the energy required for their synthesis and processing. The development of regolith-based building materials is one such area of interest, offering the potential for in-situ resource utilization to minimize dependency on Earth-based supplies.

The selection of chemicals and materials for use in space habitats is governed by an integrative assessment of mechanical properties, radiation resistance, outgassing behavior, chemical stability, compatibility with fabrication technologies, and sustainability. The relentless pursuit of advancements in material science and technology is vital to meet these criteria, aiming to ensure the safety, efficiency, and longevity of space habitats. Further developments in this field will continue to be driven by the need to address the unique operational challenges presented by the space environment.

8.4 Influence of Space Environment on Material Degradation

Material degradation in space represents a crucial consideration for the engineering and selection of substances used in constructing and maintaining space habitats. The space environment imposes a unique set of challenges including exposure to extreme ultraviolet (UV) radiation, vacuum conditions, thermal cycling, and atomic oxygen, which can severely impair the structural and functional integrity of materials.

Ultraviolet Radiation: In the absence of Earth's atmospheric protection, materials in space are directly exposed to intense UV radiation. This radiation can break chemical bonds, leading to the degradation of polymers and other organic compounds. For instance, the polyethylene used in space suits and inflatable modules undergoes yellowing and embrittlement under UV exposure. This necessitates the use of UV stabilizers within material compositions or protective coatings that absorb or reflect UV rays.

Vacuum Conditions: The vacuum of space leads to outgassing of materials, a process where volatile components within a material sublimate and escape into the environment. Common effects of vacuum-induced outgassing include the weakening of material structures and potential contamination of surrounding surfaces which can impair instrument functionality. For example, silicone rubbers frequently used for seals and gaskets are known for their high outgassing rates unless specially treated or replaced with less volatile materials.

Thermal Cycling: Space structures are subject to extreme temperature fluctuations, known as thermal cycling, ranging from intensely hot when directly exposed to the sun, to extremely cold when in the Earth's shadow. These fluctuations cause repeated expansion and contraction of materials, which can lead to fatigue and eventual fracturing. Thermal expansion coefficients must be carefully considered when designing joints and interfaces in space structures to accommodate and withstand these cycles.

Atomic Oxygen: In low Earth orbit (LEO), atomic oxygen is an aggressive oxidizer of materials, particularly affecting those exposed on the exterior of spacecraft. It effectively erodes surfaces through the oxidation process. This erosion can lead to significant material loss over time, affecting the material's mechanical properties and life span. As an example, silver-coated Teflon, used extensively for thermal control blankets, experiences significant degradation due to oxidation by atomic oxygen, thereby reducing its thermal reflectivity and mechanical durability.

Materials employed in space must, therefore, be specifically tailored or shielded to withstand these harsh conditions. Innovations in material science, such as the development of multi-layer insulation (MLI) blankets that incorporate layers of metalized polymer films and spacer fabrics, offer improved resistance against these environmental factors. Additionally, surface coatings enriched with nanoparticles have shown potential in enhancing the durability and resistance of space materials to UV radiation and atomic oxygen.

The correct understanding and anticipation of material degradation mechanisms under space environmental conditions are paramount to ensure the longevity and safety of space habitats. This insight drives not only the selection of materials but also the design of the entire habitat structure. Advances in these areas continue to be critical as missions aim for longer durations and greater distances from Earth.

In summary, addressing the challenges posed by the space environment on material degradation requires a multifaceted approach involving the selection of appropriate materials, utilization of protective coatings, and innovative design adaptations. These strategies collectively enhance the resilience of space habitats, which is vital for the successful extension of human presence beyond Earth.

8.5 Radiation Shielding Materials for Space Habitats

Radiation in space presents one of the most significant challenges for the safety and viability of habitats beyond Earth's atmosphere. Unlike Earth, which is protected by its magnetic field and atmosphere, space habitats are exposed to a spectrum of harmful radiation, including solar particles (protons) and cosmic rays (high-energy ions from outside the solar system). This radiation can penetrate habitats, damaging structural materials, and impairing the health of astronauts by increasing the risk of cancer and other radiation-induced diseases. As such, the development and application of effective radiation shielding materials is a crucial area of research for sustainable space exploration.

Materials currently employed in radiation protection for space habitats primarily function by providing physical barriers that absorb or deflect the incoming radiation. Traditional materials used include aluminum, which is favored for its structural properties and relatively good radiation shielding capabilities. However, research has highlighted significant drawbacks in using aluminum alone, primarily due to its limited effectiveness in blocking high-energy particles and the heavy weight that contributes to increased launch costs.

In search of better alternatives, recent advancements have focused on the use of high-density and hydrogen-rich materials. Hydrogen is particularly effective at scattering protons and neutrons, thereby reducing the penetration of radiation. Materials such as polyethylene and specialized hydrogenated boron nitride nanotubes (BNNTs) have emerged as promising candidates. Polyethylene, a lightweight, flexible plastic, is rich in hydrogen and has demonstrated superior performance in shielding against both solar and cosmic radiation compared to aluminum. BNNTs, embodying boron (a neutron absorber) and a hydrogenous structure, are not only efficient at shielding radiation but also possess exceptional mechanical properties, making them ideal for structural applications in space habitats.

Moreover, researchers are exploring the innovative concept of multifunctional materials that integrate radiation shielding with other critical functionalities, such as structural integrity and thermal regulation. Composite materials combining metals, polymers, and in some cases, microencapsulated phase change materials, are under investigation. These composites aim to not only provide effective radiation protection but also to contribute to the thermal management of the habitat and to mitigate against other environmental challenges of space.

Another intriguing development in radiation shielding is the utilization of regolith, the layer of loose, heterogeneous material covering solid rock, found on the Moon and Mars. Techniques are being developed to use in-situ resource utilization (ISRU) to convert regolith into building materials through processes like sintering, melting, or using it as aggregate in composite materials. This approach not only reduces the need for transport of materials from Earth but also provides a use for readily available local resources.

The selection and development of radiation shielding materials for space habitats involve a complex balance of factors, including effectiveness, weight, dual functionality, and feasibility of production. As space travel extends to more distant locations such as Mars, these materials will play a critical role in ensuring the safety and success of human space exploration. At present, while materials like polyethylene and BNNTs show considerable promise, ongoing research and testing are essential to optimize these materials for actual deployment in space habitats. The continuing evolution of material science, therefore, remains fundamental to overcoming the challenges posed by the space environment.

8.6 Self-Healing Materials for Long-Duration Missions

The integrity of materials used in space habitats is paramount, especially under the harsh conditions found in extraterrestrial environments. Considering the logistical and economic challenges associated with transporting materials from Earth to space, it becomes critical to develop materials that can autonomously repair damage incurred during long-duration missions. This specific section delves into the science of self-healing materials, exploring their chemistry, mechanisms, and potential applications in space exploration.

Self-healing materials are engineered to automatically restore their structure and function after suffering physical or mechanical damage. These materials mimic biological systems such as human skin, which heals after an injury. For instance, materials embedded with microcapsules containing a healing agent that can rupture and release upon damage are typical examples. When a crack forms, the microcapsules break open, and the healing agent reacts chemically with a catalyst embedded within the material matrix, resulting in polymerization or curing that heals the crack.

The chemistry behind self-healing materials often involves polymerization reactions. For example, an epoxy matrix imbued with a latent hardener and a microencapsulated healing agent can effectively restore its structural integrity post-damage. In response to microcracking, the microcapsules spill their content into the crack, where the healing agent reacts with the hardener to polymerize and solidify, rebonding the crack.

In space applications, these technologies are invaluable. The vacuum of space, combined with microgravity, radiation, and temperature extremes, can exacerbate material degradation. Self-healing materials can obviate the need for frequent extravehicular activities to inspect and repair exterior surfaces of spacecraft and habitats, reducing risks to astronauts and lowering mission costs.

A remarkable application of these materials is seen in the experimental use on the International Space Station (ISS). The ISS employed a self-healing polymer that could recover its integrity after small-scale meteorite impacts. This capability is especially critical given the high velocity of space debris which may impact the station.

Another promising development is the integration of self-healing materials with additive manufacturing techniques such as 3D printing. This integration is particularly suitable for constructing habitats on the Moon or Mars, where materials with high durability and low maintenance are advantageous. Researchers have been exploring the incorporation of self-healing agents into 3D printable materials, aiming to develop structures that can autonomously repair damages caused by environmental factors or human activities.

Furthermore, the scientific community is investigating the possibility of leveraging the unique properties of certain local Martian and lunar materials. For example, sulfur-based compounds found abundantly on Mars show promise as binders in the construction of Martian concrete. These materials could be crafted to possess self-healing properties by formulating them with appropriate polymeric agents that allow the material to maintain functionality in the Martian environment.

Self-healing materials hold a pivotal role in the sustainability and safety of extended space missions. By reducing the need for manual repairs and enhancing the longevity of critical infrastructure, these innovative materials not only promise to protect costly equipment and precious human lives but also herald a new era in the design and construction of extra-terrestrial habitats, where sustainability and efficiency are paramount. Continued research and development into the mechanisms and applications of self-healing materials, along with their integration into existing space technology, will be crucial in advancing our capabilities for long-term space exploration.

8.7 Sustainable Production of Materials in Space

The quest for sustainability in space missions is critical, particularly as the duration of missions extends and plans for permanent off-Earth habitats progress. Sustainable production of materials in space environments focuses on minimizing waste, reducing reliance on Earth-supplied materials, and maximizing the efficiency of resource usage. This involves integrating advanced technologies and innovative approaches, such as closed-loop systems, in-situ resource utilization (ISRU), and regenerative life support systems.

Closed-loop systems are essential in the sustainable production of materials, ensuring that all resources are recycled and reused. In space habitats, closed-loop systems convert waste products back into usable resources. For example, water recovery systems onboard the International Space Station (ISS) reclaim over 90% of water from urine, sweat, and exhaled moisture. This system involves several stages of filtration and treatment to provide clean water for drinking and other uses, demonstrating how cycle-based systems can significantly contribute to sustainability in constrained environments.

In-situ resource utilization (ISRU) is another cornerstone for sustainable material production, focusing on the use of local resources to create necessary materials and support systems. This technique reduces the need for costly and logistically challenging resupply missions from Earth. On the Moon and Mars, ISRU can utilize abundant local materials such as regolith. The regolith can be processed into building materials for habitats. One promising method involves sintering regolith using microwave energy to produce solid bricks. This technique not only provides efficient material production without the need for transporting conventional building materials but also offers the potential for significant structural integrity suitable for protective constructions against environmental extremes.

Furthermore, photocatalytic materials have been developed to split water into hydrogen and oxygen using solar energy. This process not only provides vital life support elements but also fuels for power generation. The integration of these materials into space habitats introduces a self-sustaining system for air and water recycling, thus contributing significantly to the sustainability of long-term space missions.

8.7. SUSTAINABLE PRODUCTION OF MATERIALS IN SPACE

Regenerative life support systems (RLSS) extend sustainability in space habitats by continuously recycling and renewing biological and material resources. An example of RLSS recently tested involves the BioRegenerative Life Support System (BLiSS), which integrates hydroponic and aquaponic methods to manage waste and produce food. These systems use plants and microorganisms to purify the air and water, while providing food to the crew, thereby creating a micro-ecosystem that mimics Earth's natural ecological cycles.

The application of 3D printing technology in space also plays a crucial role in sustainable material production. Using resources such as recycled plastics or processed regolith, 3D printers can produce a variety of tools, components, and even habitat structures on demand, reducing the mass, waste, and inefficiencies associated with transporting goods from Earth. For instance, a 3D printer on the ISS has successfully manufactured items from recycled materials, showing the potential of additive manufacturing in resource-limited settings.

The transitions toward sustainable production in space are not merely technological challenges but also involve economic and policy considerations. Developing comprehensive, cross-disciplinary approaches that incorporate engineering, biological and physical sciences, economics, and policy planning are essential for the realization of sustainable extraterrestrial environments.

Given these advanced technologies and strategies for sustainable material production, future space missions can look forward to more self-sufficient and ecologically responsible operations, paving the way for longer and potentially indefinite stays in outer space.

Problems

- Discuss the role of closed-loop systems in sustainable material production in space habitats and provide examples of such systems currently in use.

- Analyze the challenges and potential solutions for implementing ISRU on Mars considering current technological developments.

- Describe how 3D printing technology can aid in sustainability on long-duration space missions. Include current examples and speculate on future developments.

- Explain the concept of regenerative life support systems and discuss their importance for future sustainable space habitats.

8.8 Chemical Processes for Air and Water Recycling

The effective recycling of air and water is paramount to the sustainability of long-duration space missions and habitats. This section explores the chemical processes integral to recycling aboard spacecraft and permanent extraterrestrial habitats, with an emphasis on molecular-level mechanisms and system design considerations.

The basic premise of air recycling in space revolves around removing carbon dioxide (CO_2) produced by astronauts and regenerating essential oxygen (O_2). One of the primary methods employed is the Sabatier Reaction, a process where CO_2 is combined with hydrogen (H_2) in the presence of a nickel catalyst at elevated temperatures to produce methane (CH_4) and water (H_2O). The overall reaction can be represented as:

$$CO_2 + 4H_2 \rightarrow CH_4 + 2H_2O.$$

The produced water can then be electrolyzed, splitting it back into oxygen and hydrogen, where oxygen is utilized for breathable air, and hydrogen is recycled back into the Sabatier process.

Another critical system is the Electrolytic Oxygen Generator (EOG), which uses the principle of water electrolysis to split water into oxygen and hydrogen by applying an electric current:

$$2H_2O(l) \rightarrow 2H_2(g) + O_2(g).$$

This process not only provides necessary oxygen but also assists in balancing humidity within the habitat.

In terms of water recycling, the Urine Processor Assembly (UPA) and the Water Processor Assembly (WPA) play essential roles. The UPA utilizes distillation — a process where urine is boiled and the vapors are condensed into liquid water, leaving behind contaminants. Subsequent treatment stages involve a series of multifiltration beds which contain substances that adsorb or chemically react with remaining impurities including organic compounds and microorganisms.

Furthermore, the WPA harnesses a combination of processes to treat all wastewater aboard the space habitat, including sweat, shower runoff, and humidity condensate. Advanced oxidation processes (AOPs) are used where powerful oxidants like hydroxyl radicals are generated in situ to break down organic contaminants into carbon dioxide, water, and simple salts. The catalysts, typically metal oxides, are crucial for effective reaction rates under the spatial constraints and power limitations inherent in space environments.

A key component to maintaining the efficiency of these recycling systems involves strict monitoring and regulation of pH levels, concentration of heavy metals, and biocontaminant levels via onboard analytical instruments. The introduction of smart sensors and automation has enhanced the reliability and autonomy of these processes, reducing the need for human intervention and allowing for real-time adjustments based on dynamic on-board environmental conditions.

The integration and interaction of these chemical processes form a closed-loop system that significantly reduces the need for resupply missions, thereby enhancing the sustainability and feasibility of prolonged human presence in space. These systems are not only critical in space but also serve as models for sustainable water and air management on Earth, particularly in isolated or resource-limited environments.

The mastery of chemical processes in the recycling of air and water is a cornerstone for advancing human activities in space. By optimizing these systems, space missions can achieve higher levels of self-sufficiency, critical not only for the health and safety of astronauts but also for the broader goal of sustainable interplanetary colonization.

8.9 Advanced Fabrication Techniques: 3D Printing in Space

The advent of three-dimensional (3D) printing technology has ushered in a transformative phase in the methods of fabricating various structures, crucially impacting the development of space habitats. The capability to construct components directly in space conditions provides numerous benefits, including minimizing the launch weight and increasing the flexibility of manufacturing mission-specific tools and spare parts on-demand.

3D printing, also known as additive manufacturing, builds objects layer by layer from a digital model. This technology uses various materials such as polymers, metals, and composites, which are selected based on the specific requirements of the application, including mechanical strength, flexibility, and resistance to extreme conditions. The precise control offered by this technology allows for the creation of complex structures which are otherwise difficult or impossible to produce with traditional manufacturing techniques.

In the context of space habitats, the utility of 3D printing lies not just in its ability to produce necessary structures and components on-demand but in its potential for resource efficiency. For instance, the ability to recycle and reprocess materials aboard spacecraft reduces the need for extensive raw material storage and minimizes waste. This is critical for long-duration missions where carrying vast amounts of construction material is impractical.

A notable application of 3D printing technology in space is the production of habitat structures on planetary surfaces such as the Moon or Mars. Research efforts like NASA's 3D-Printed Habitat Challenge have explored the possibility of using regolith, the loose rock and dust on the planetary surface, as a building material. By combining a binding agent with regolith, robust structures suitable for habitation can be printed that withstand the harsh conditions of space environments.

Additionally, in zero-gravity conditions, 3D printing must adapt to the absence of a supporting structure which on Earth assists in layering the material continuously. Innovations have led to the development of specialized printers capable of operating in microgravity conditions aboard the International Space Station (ISS). The European Space Agency (ESA) and other international bodies have successfully conducted experiments with bio-printing in space, focusing on the construction of human tissue and fusing severed connections.

Metals such as titanium and steel are used for printing components of spacecraft because of their strength and durability. These materials can be printed using laser sintering methods, which involve lasers that accurately melt the metal powders to form solid, durable objects. Laser sintering in the space environment needs to be carefully controlled to manage the rapid cooling rates and the unique behaviors of molten metals in low gravity.

Furthermore, the integration of sensors and electronic components directly into 3D printed parts is under exploration. This could lead to the next generation of smart space devices with in-built monitoring systems, significantly enhancing functionality and reliability.

The implementation of 3D printing technologies in space operations reflects a shift towards more adaptive, economically feasible space missions that are crucial for the long-term exploration and colonization of outer space locations. As research progresses, it will not only become a standard practice within astronauts' skillsets but also evolve the scope of space missions by providing sustainable options for habitation and operations.

To corroborate the theoretical advantages and understand the real-world implications, mathematical problems related to the heat diffusion rates during metal sintering processes, optimization algorithms for minimum material usage while maintaining structural integrity, and scenarios involving the redesign of structural components to suit unexpected changes in mission parameters should be explored deeply by students. Thus, reinforcing their grasp of the necessary chemistry and material science principles involved in the advanced, yet practical scenario of 3D printing in space habitats.

8.10 Case Studies: Material Use in the ISS and Mars Habitats

The International Space Station (ISS) and envisioned Martian habitats exemplify the practical application and challenges of selecting appropriate materials and chemicals in a space environment. In this section, we will explore and analyze the material utilization strategies employed in the ISS, and how these strategies pave the way for future Mars habitats.

The ISS, orbiting Earth since 1998, serves as a microgravity and space environment research laboratory where astrochemistry, astronomy, meteorology, physics, and other fields converge. Its construction and maintenance have required materials that can withstand the harsh conditions of space, including extreme variations in temperature, micrometeoroid impacts, and high levels of cosmic and solar radiation. Among the most crucial materials used are high-strength aluminum alloys for the main modules, which provide a balance between strength, ductility, and corrosion resistance essential for structural integrity. These aluminum alloys are often coupled with thermal protection systems that include multi-layer insulation—reflective sheets interspersed with insulating material to shield against the severe thermal fluctuations between sunlight and shade in orbit.

Solar panels are another critical component, made from layers of silicon cells that must not only be efficient but also resistant to radiation degradation over extended periods. To boost their longevity, recent ISS upgrades have adopted newer technologies such as layered III-V compound semiconductors, which include materials like gallium arsenide that provide better efficiency and radiation hardness than traditional silicon cells.

Similar material considerations have guided the development strategies for Mars habitats. Martian conditions—characterized by subsurface temperatures averaging about -63 degrees Celsius, thin carbon dioxide-rich atmosphere, and frequent dust storms—demand materials that can provide insulation, sustainability, and self-healing properties. Innovations are increasingly geared toward polymers and composites that can be synthesized on Mars from local resources. For example, research into using basalt, a common Martian rock, to produce fibers for construction materials is underway. These fibers can potentially be used to reinforce structures similarly to how steel reinforces concrete on Earth.

Furthermore, the development of habitat structures able to withstand Martian dust storms has directed attention to the use of aerogel composites. These materials, known for their lightweight and high insulation properties, prove ideal in maintaining internal temperatures and protecting from external elements. Aerogels are being considered for inclusion in the walls of Mars dwellings, interwoven with other composites derived from in-situ resources.

One of the most ambitious material technologies being explored is the use of 3D printing techniques to construct habitats from regolith—the loose rock and dust covering the planet's surface. This would reduce the need to transport heavy building materials across interplanetary space, significantly cutting the cost and complexity of Mars missions. The utilization of regolith has been demonstrated successfully in prototype structures on Earth, where simulated Martian soil has been processed into a durable concrete-like material using a binding agent that could potentially be manufactured on Mars.

Transitioning from the ISS to Mars habitats involves not only adapting existing materials but also pioneering new technologies. For instance, radiation shielding remains a profound challenge on Mars, much more so than on the ISS. The development of regolith-based

8.10. CASE STUDIES: MATERIAL USE IN THE ISS AND MARS HABITATS

construction materials also incorporates elements such as hydrogen-rich compounds to enhance their radiation-absorbing capabilities, which is crucial for long-term human habitation.

This exploration into material usage in the ISS and Mars habitats illustrates a gradual but innovative transition from utilizing Earth-sourced, high-performance materials to developing sustainable, self-sufficient production capabilities using Martian resources. The progression in material technology from the ISS to proposed Mars habitats highlights not only the evolution of material science but also humanity's growing capability to adapt and thrive in extraterrestrial environments. Through these case studies, it becomes evident that the future of space exploration deeply intertwines with advancements in chemical and material sciences, paving the way for the next leaps in human space habitation.

CHAPTER 8. MATERIALS AND CHEMICALS FOR SPACE HABITATS

Chapter 9

Chemical Sensors and Instruments for Space Missions

This chapter outlines the development and implementation of chemical sensors and analytical instruments essential for space missions. It covers the various types of sensors and instruments used to detect and analyze chemical species in diverse space environments, from planetary atmospheres to deep space. The text highlights the technological innovations that enhance sensitivity and specificity, address the challenges of operating in harsh conditions, and contribute to our understanding of celestial chemistry. Additionally, it discusses how these tools are integral to studies ranging from environmental monitoring aboard spacecraft to exploration and sample analysis on alien worlds.

9.1 Overview of Chemical Sensing in Space Exploration

Chemical sensing in space exploration is a pivotal aspect of understanding and interacting with extraterrestrial environments. The ability to detect and analyze chemical species with a high level of precision allows scientists to draw significant conclusions about the composition, history, and potential habitability of other planets, moons, and celestial bodies. This section aims to elucidate the critical role that chemical sensors play in the endeavor of space exploration, delineating their functioning, importance, and the scientific insights they facilitate.

The essence of chemical sensors for space missions is rooted in their capability to provide accurate and real-time data regarding the chemical environment of space. These sensors are meticulously designed to withstand the severe conditions of space, which range from extreme temperatures to radiation levels that would compromise most standard instrumentation. Chemical sensors employed in space missions are essentially the ambassadors of Earth's scientific community, reaching far into the cosmos to retrieve and relay data that would otherwise be beyond our grasp.

From the Mars rovers that analyse soil samples to the spectrometers aboard the Hubble Space Telescope, a broad array of instruments serves as the backbone of chemical sensing in space. Each tool is precisely engineered to target specific chemical signatures. For instance, spectrometers measure the properties of light to deduce the composition of planetary atmospheres, while gas chromatography systems separate and identify gases and volatile compounds.

The operational parameters of these sensors are largely dictated by the extreme space conditions they must endure. For example, instruments on the surface of Mars have to function effectively within both the thin carbon dioxide-rich atmosphere and the wide temperature fluctuations, which can range from 20 degrees Celsius at noon to below -73 degrees Celsius at night. Such instruments are not only rugged but are also highly autonomous, capable of making adjustments based on the environmental data they themselves collect.

Moreover, chemical sensors help to discern vital information about potential life supporting conditions by identifying water presence, organic compounds, and other life-supporting chemicals. The famous detection of methane spikes on Mars by NASA's Curiosity rover stands as a quintessential example, where periodic increases in methane levels hinted at possible microbial activity, or perhaps interactions between water and rock.

The data gathered through these sensors not only enrich our understanding of the chemical properties and processes in space but also enhance our ability to predict and model conditions in extraterrestrial environments. This capability is crucial for future manned missions and potential colonization as it provides necessary information that ensures the safety and sustainability of human presence on other planets.

In sum, the role of chemical sensing in space exploration is fundamental and all-encompassing. These sensors and instruments are key to deciphering the complexities of space environments. They offer factual evidence that guides scientific theories and enhances the breadth and depth of human knowledge regarding the universe. Moving forward, innovations and advancements in this field will continue to illuminate the unknown, paving the way for further exploration and perhaps, one day, human habitation beyond Earth.

9.2 Fundamental Principles of Chemical Sensors

Chemical sensors are devices that translate chemical information, ranging from the concentration of a specific sample component to total composition analysis, into analytically useful signals. The fundamental principles governing the operation of chemical sensors are selectivity, sensitivity, stability, and reproducibility, which are essential for their effective application in the rigorous confines of space missions.

Selectivity in chemical sensors refers to the ability of a sensor to distinguish between different chemical species in a mixture. In the context of space exploration, selectivity is crucial due to the complex compositions of planetary atmospheres and extraterrestrial surfaces. The specific interaction between a sensing material and the target analyte, facilitated through physical adsorption or chemical bonding, enables this selectivity. Examples of materials with high selectivity include metal oxides for gas sensing and polymer coatings for volatile organic compounds in spectroscopy.

Sensitivity, the second principle, measures the minimum amount of substance necessary to produce a measurable response from the sensor. The sensitivity of a sensor is typically enhanced by maximizing the area of interaction between the target analyte and the sensor's active site. In space applications, sensitivity must be sufficient to detect extremely low concentrations of analytes, given the dilute nature of extraterrestrial environments. Techniques such as amplifying electronic signals generated by the interaction or increasing the surface area of the sensor material through nanostructuring are commonly used to boost sensitivity.

Stability, another critical factor, represents the degree to which a sensor maintains its performance over time under varying environmental conditions. The harsh conditions of space—extreme temperatures, radiation, and vacuum conditions—demand that chemical sensors not only initially meet performance standards but also maintain their functionality over long durations. Materials that resist degradation or changes in physical properties when exposed to environmental stresses are preferred. Examples include inert metals and stable oxides, which can provide reliable performance without substantial drift in sensor readings.

Reproducibility in sensor responses ensures that the sensor gives consistent readings when exposed to the same conditions on different occasions. This attribute is vital in space missions where differing results can lead to confusion and misinterpretation of environmental data. Reproducible results are achieved through careful control of the manufacturing process, ensuring sensor materials and structures are consistent from one device to another.

Advancements in microfabrication and nanotechnology have played a pivotal role in enhancing the performance of chemical sensors by improving these fundamental attributes. For instance, employing microelectromechanical systems (MEMS) technology allows for the development of smaller, more energy-efficient sensors that are capable of real-time monitoring and analysis. These miniaturized sensors are easily integrated into various spacecraft components, aiding in continuous monitoring without significantly adding to payload weight.

The effectiveness of chemical sensors in space exploration hinges significantly on these fundamental principles. Enhanced selectivity, sensitivity, stability, and reproducibility not only ensure accurate data collection and analysis but also contribute to the overall success of a mission by providing reliable monitoring of chemical bearing environments encountered in space.

9.3 Types of Chemical Sensors Used in Space Missions

Chemical sensors play a pivotal role in the exploration and study of space. These sensors are designed to detect and analyze the chemical composition of their environments, producing data crucial for astronomical studies, planetary science, and the safety of space missions. We classify these sensors based on their detection principles and the types of measurements they provide. This section focuses on electrochemical sensors, optical sensors, and mass spectrometers as they are widely incorporated in various space missions.

Electrochemical Sensors: Electrochemical sensors operate based on the measurement of electrical parameters that change in response to chemical reactions. These sensors are particularly useful for their sensitivity to a wide range of analytes, low power requirements, and compact size, which are essential attributes for space equipment. A common application is the detection of atmospheric gases on manned spacecraft and space stations, such as carbon dioxide and ammonia, both of which pose significant risks to air quality and human health. Another example is the use of potentiometric sensors to measure the pH of fluids in life support systems, ensuring that such systems are safe and effective for crew members.

Optical Sensors: These sensors utilize the interaction of light with matter to detect changes in chemical composition. Their design includes components such as lasers, photodetectors, and optical fibers. Optical sensors are highly sensitive and capable of providing real-time analysis, which makes them invaluable for detecting trace levels of gases and other substances. A notable utilisation is in the monitoring of molecular oxygen and water vapor, critical for both life support systems and scientific experiments. Moreover, optical sensors based on fluorescence spectroscopy have been developed for remote sensing applications on planetary rovers, aiding in the mineralogical examination of Martian and lunar surfaces.

Mass Spectrometers: Mass spectrometry is a powerful analytical technique used in space missions for its ability to identify unknown compounds and quantify known materials with high accuracy. Spaceborne mass spectrometers have been integral in missions such as the Mars Curiosity Rover, analyzing rock samples to assess past habitability. These instruments work by ionizing chemical species from a sample, sorting the ions based on their mass-to-charge ratios, and detecting them to provide a mass spectrum. Each peak in this spectrum corresponds to a chemical component, enabling detailed compositional analysis of solid, liquid, and gaseous samples.

Semiconductor Sensors: Particularly useful in detecting gases, semiconductor sensors operate by changes in electrical resistance across a semiconductive material as it interacts with an analyte. These sensors are compact, robust, and have a high degree of sensitivity and selectivity. The Mars Science Laboratory employs semiconductor sensors to monitor methane — a gas of significant interest due to its potential implications of biological processes on Mars.

The variety of chemical sensors utilized in space missions reflects the diverse requirements and challenges of chemical analysis in extreme environments. These sensors provide critical data that help understand extraterrestrial chemistry, assess planetary habitability, and ensure the safety and success of human activities in space. The continuous evolution of these technologies is essential to the advancement of space exploration, pushing the boundaries of what can be detected and analyzed in the harsh conditions of space.

9.4 Role of Spectroscopy in Chemical Analysis

Spectroscopy, a fundamental analytical technique, plays a pivotal role in the chemical analysis of materials encountered during space missions. Its utility lies in its ability to provide detailed information about the composition, structure, and physical properties of unknown substances without requiring direct contact. This section elucidates the various spectroscopic methods employed in space exploration, their working principles, and their applications in the analysis of celestial bodies.

At the heart of spectroscopy is the interaction between electromagnetic radiation and matter. When electromagnetic radiation—be it visible light, ultraviolet, infrared, or X-rays—impinges on a sample, it can be absorbed, emitted, or scattered, depending on the nature of the sample. These interactions yield a spectrum, which is essentially a fingerprint unique to each material. By analyzing these spectral fingerprints, scientists can deduce various chemical and physical properties of the matter under study.

9.4. ROLE OF SPECTROSCOPY IN CHEMICAL ANALYSIS

Among the most commonly used spectroscopic techniques in space missions are absorption spectroscopy, emission spectroscopy, and Raman spectroscopy. Absorption spectroscopy measures the wavelengths of light absorbed by a sample. This method is incredibly useful for determining the molecular identity and concentration of substances in planetary atmospheres. For instance, Mars missions utilize infrared absorption spectroscopy to detect water vapor and carbon dioxide levels, providing invaluable data for climatic and geological studies.

Emission spectroscopy, on the other hand, involves analyzing the light emitted by a substance. When a sample is excited, typically by heating or by exposing it to light, it emits light as it returns to its ground state. The emission spectrum can help in identifying elements and compounds with a high degree of specificity and sensitivity. This technique has been crucial in studying the mineral composition and surface features of the Moon and other celestial bodies, aiding in the mapping of their geochemical properties.

Raman spectroscopy, which observes the scattering of light, offers unique advantages in space explorations, such as minimal sample preparation and resistance to interference from water molecules. It is particularly advantageous for robotic missions where equipment needs to be autonomous and robust. Raman spectrometers have been deployed on Mars rovers to analyze rock formations, detect organic compounds, and further our understanding of Mars' geology.

Moreover, advancements in spectroscopic technology have led to the development of compact, efficient instruments suitable for space applications. Miniaturized spectrometers, for example, have been integrated into space telescopes and rovers, expanding the scope of scientific experiments that can be conducted remotely. These instruments often incorporate hybrid techniques, such as combining mass spectrometry and gas chromatography with traditional spectroscopy, to enhance analytical capabilities and achieve comprehensive chemical profiling.

The data collected through these spectroscopic analyses are pivotal not only in answering fundamental astrobiological questions, such as the potential for life on other planets, but also in practical applications such as assessing potential hazards or resources for future human missions. They continue to shape our strategies for exploration and our understanding of space environments.

In practical application terms, spectroscopy's role extends beyond mere identification. It can also quantify the abundance of elements and compounds, monitor dynamic changes over time, and provide data for models of planetary atmospheres and surfaces. Thus, it is an indispensable tool in the arsenal of techniques used for chemical analysis in space exploration.

9.5 Mass Spectrometry Instruments for Space Applications

Mass spectrometry (MS) has become an indispensable tool in the arsenal of techniques employed for chemical analysis in space exploration. This technique relies on the ionization and characterization of chemical species based on their mass-to-charge ratio (m/z). The inherent sensitivity and specificity of MS make it an excellent choice for the analysis of complex mixtures in extraterrestrial environments, where small sample sizes and the potential for unknown compounds are significant challenges.

The application of mass spectrometry in space missions primarily revolves around its capability to detect and quantify organic molecules, monitor potential life markers, and analyze the composition of planetary atmospheres and surfaces. Advanced MS instruments have been designed to operate under the vacuum and temperature extremes typical of space environments, and they often incorporate miniaturization and ruggedization strategies to meet the strict payload constraints of spacecraft.

One of the pivotal developments in this area has been the use of Time-of-Flight (ToF) mass spectrometry. Instruments based on the ToF principle, such as the Mars Organic Molecule Analyzer (MOMA) aboard the ExoMars rover, utilize a swift and effective method to analyze the chemical composition of Martian soil. ToF-MS instruments record the flight time of ionized atoms or molecules from the source to the detector; each species' time of flight is indicative of its mass-to-charge ratio. This methodology is particularly well-suited for space applications due to its rapid analysis capability, which is vital for the sometimes limited operational time frames on missions.

Moreover, ion trap mass spectrometers, such as those aboard the Cassini spacecraft, have conducted in-depth studies of the atmospheres and rings of Saturn. These instruments capture and store ions in a magnetic or electric field, allowing for extended analysis which is crucial in obtaining a detailed compositional profile. The data provided by these spectrometers have deepened our understanding of the chemical interactions occurring in the outer solar system and provided insights into cosmological processes.

Beyond planetary research, mass spectrometry is also instrumental in the characterization of cometary and interstellar samples. The Rosetta mission's Rosetta Orbiter Spectrometer for Ion and Neutral Analysis (ROSINA) was notably equipped with a Double Focusing Mass Spectrometer (DFMS) and a Reflectron type Time-of-Flight (RTOF) spectrometer, which played a key role in analyzing the comet 67P/Churyumov-Gerasimenko. This mission provided an unprecedented glimpse into the molecular complexities of a comet, revealing a wealth of organic compounds and suggesting mechanisms of prebiotic chemistry in early solar systems.

Given the complex nature of space missions, the integration of MS instruments often requires interdisciplinary collaboration. Calibration against terrestrial standards, validation under simulated environmental conditions, and comprehensive data analysis frameworks are essential components in the utilization of MS data for space research. Additionally, the future of MS in space exploration looks toward even greater miniaturization, enhanced sensitivity, and perhaps even the integration of artificial intelligence to handle vast amounts of spectral data autonomously.

As the exploration of space progresses, the role of mass spectrometry in unlocking the mysteries of the cosmos only deepens. Continuing technological advancements are expanding the potential of what we can learn from the smallest particles collected from the furthest reaches of space, reaffirming the importance of precise and reliable instruments in the ongoing quest to understand our universe.

9.6 Gas Chromatography for Analyzing Planetary Atmospheres

Gas Chromatography (GC) has emerged as a pivotal analytical technique in the study of planetary atmospheres within the realm of space exploration. Employed to separate and analyze compounds that can be vaporized without decomposition, GC is particularly suited for the molecular analysis of extraterrestrial atmospheres, which frequently contain complex mixtures of organic and inorganic volatiles.

The principle of GC involves the vaporization of the sample followed by its passage through a long, thin column coated with a stationary phase. The individual components of the sample are separated based on their differential interactions with the stationary phase and their respective volatilities. A carrier gas, often helium due to its inert properties, enables the transfer of vaporized molecules through the column. As molecules exit the column at different times (retention time), they are detected and quantified, usually by a detector that responds to the amount of substance eluting from the column per unit time.

In the context of space missions, GC systems are designed to be compact, robust, and autonomously operational. The Mars Science Laboratory's Curiosity Rover, for instance, employs a miniaturized GC system as part of the Sample Analysis at Mars (SAM) instrument suite. SAM's GC system analyzes gases evolved from rock and soil samples heated in an oven, allowing for the identification of organic compounds that might suggest past or present life.

Detecting atmospheric constituents in locations such as Titan, the largest moon of Saturn, illustrates further the indispensable role of GC in planetary exploration. The Gas Chromatograph Mass Spectrometer (GC-MS) aboard the Huygens probe was specifically engineered to analyze the chemical composition of Titan's atmosphere, which is thick with nitrogen and methane. The GC-MS data revealed the presence of complex hydrocarbons, expanding our understanding of atmospheric processes in environments radically different from Earth.

The adaptability of GC to various detection approaches enhances its utility for space applications. Typically, thermal conductivity detectors (TCD) and flame ionization detectors (FID) are employed, with TCD being preferred for inorganic gases and FID for organic analysis. However, for space missions where power, size, and sensitivity are crucial, more sophisticated detection methods such as mass spectrometry are often integrated with GC systems to create GC-MS technology. This combination provides high sensitivity and selective detection capability essential for deciphering the complex chemical environments of extraterrestrial atmospheres.

From a design perspective, the challenges of implementing GC in space missions include miniaturization, resistance to extreme environmental conditions such as temperature fluctuations and cosmic radiation, and the capability for autonomous operation with minimal maintenance. Innovations in materials science have led to the development of more robust, chemically inert column coatings and the utilization of advanced microfabrication techniques for constructing compact, efficient GC systems.

Future missions might see enhanced applications of GC, facilitated by advancements in microfluidic technologies and nanomaterials. These developments could lead to even more sensitive and selective sensors capable of detecting a broader range of chemical species at lower concentrations. Furthermore, the integration of AI and machine learning for real-time data analysis could significantly optimize the detection and interpretation of atmospheric data captured by GC, pushing the boundaries of our chemical understanding of other worlds.

By contributing to a detailed molecular understanding of planetary atmospheres, gas chromatography stands as a cornerstone of chemical analysis in space exploration. Through ongoing advancements in GC technology and methodology, scientists continue to refine our understanding of the cosmos, leading to more informed decisions about future missions and the potential habitability of other planets.

9.7 Developments in Nanochemical Sensors for Space

The evolution of nanochemical sensors marks a significant revolution in space exploration technologies, primarily driven by the need for highly sensitive, robust, and miniaturized devices capable of operating within the harsh environments of space. Nanochemical sensors, characterized by their nanometric scale functional elements, offer unprecedented advantages in terms of sensitivity, specificity, and rapidity of response, which are crucial for real-time analysis in space missions.

The principle behind nanochemical sensors involves the utilization of nanomaterials, such as carbon nanotubes, graphene, metallic nanoparticles, and semiconductor quantum dots, which exhibit unique electrochemical, optical, and mechanical properties not found in bulk materials. These properties arise from the quantum size effects and the high surface-to-volume ratio of nanomaterials, enhancing their interaction with analyte molecules. This intense interaction leads to more significant changes in physical properties upon exposure to target analytes, thus enabling the detection and quantification of chemical species at extremely low concentrations.

Significant advancements have been made in the development of nanostructured metal oxides for use as sensitive layers in chemical sensors. These materials are particularly effective in detecting atmospheric gases and volatile organic compounds (VOCs), essential for monitoring spacecraft air quality and analyzing extraterrestrial atmospheres. Metal oxide nanostructures, when exposed to target gases, undergo changes in their electrical resistance, which can be measured and related to the gas concentration. The high reactivity and fast response times of these nanostructured materials are due to the abundant reactive sites and short diffusion paths for gases.

Another noteworthy development in nanochemical sensing technology is the creation of nanosensor arrays, inspired by the biological olfactory systems. These arrays consist of multiple sensing elements, each tailored to respond distinctly to different chemicals. By analyzing the pattern of responses from the entire array, complex mixtures of chemicals can be identified and quantified. This technology mimics the human nose's ability to perceive and distinguish among different odors and has been crucial for the analysis of planetary atmospheres where the chemical environment is complex and largely unknown.

Furthermore, the integration of nanochemical sensors with microelectronic devices has led to the development of smart sensor systems that can operate autonomously, perform on-site data processing, and communicate results back to Earth or to local spacecraft systems. These integrated devices are designed with power efficiency and long-term stability in mind to withstand the long durations and extreme conditions of space missions.

For instance, recent missions have employed nanochemical sensors to detect water vapor and potential biomarkers on Mars, demonstrating their practical utility and reliability. The sensors used on these missions are a testament to the significant strides made in this field, showing not only the capability to detect the presence of specific analytes but also to operate effectively in the fluctuating temperatures and radiation levels characteristic of space environments.

The developments in nanochemical sensors for space application represent a pivotal leap forward in our ability to explore and understand extraterrestrial environments. These sensors are crucial for environmental monitoring, life-support systems in manned missions, and astrochemical studies. They continue to evolve, pushing the boundaries of what can be detected and analyzed in the vast reaches of space. As the technology matures and more exploratory missions are launched, nanochemical sensors are expected to play an increasingly central role in unraveling the chemical mysteries of the universe.

9.8 Integration of Sensors in Robotic and Human Missions

The integration of chemical sensors in space exploration missions, whether robotic or manned, plays a pivotal role in enhancing the scientific yield and ensuring the safety of the missions. Chemical sensors are intricately designed to operate within the complex systems of spacecraft and space suits, providing critical data for environmental control, life support, and scientific experiments.

In robotic missions, sensors are primarily tasked with environmental analysis and sample collection. A quintessential example is the Mars Science Laboratory rover, Curiosity, which employs a suite of sensors capable of analyzing soil and atmospheric samples. These sensors detect the presence of various gases like methane, which has crucial implications for understanding potential life on Mars. The integration of these sensors involves meticulous planning to ensure their functionality and reliability in the harsh Martian environment, which includes extreme temperatures and high radiation levels.

The design of these sensors often requires miniaturization and ruggedization. Miniaturization allows the inclusion of multiple sensors without significantly increasing the payload, while ruggedization ensures that sensors can withstand the vibrations during launch and landing as well as the abrasive particulates found on planetary surfaces. Moreover, robotic missions also utilize remote sensing technology to perform spectroscopic studies from orbit, necessitating the integration of spectrometers that can operate across various wavelengths to identify chemical signatures.

For human missions, chemical sensors are integrated not only for scientific investigation but also to monitor the health and safety of astronauts. For example, sensors in the life support system of the International Space Station analyze the air for contaminants like carbon dioxide and ammonia, ensuring that the atmosphere is within safe breathing parameters. These sensors must deliver real-time results to enable quick responses to potential life-threatening anomalies.

The integration process in manned missions extends to the development of portable and wearable sensors that astronauts can use during extravehicular activities. These devices monitor external environmental conditions and provide telemetry back to the mission control about the chemical composition found in unexplored areas. They are designed to be operable with space gloves and resistant to the ultraviolet radiation pervasive in space.

In both robotic and human missions, data from chemical sensors are crucial for the operation of closed-loop systems designed for environmental control and life support. These systems rely on sensors to adjust parameters dynamically in response to changing environmental conditions or the metabolic activity of astronauts. The integration of sensors thus not only supports scientific objectives but also underpins the adaptability and sustainability of life support systems.

The strategic placement of sensors throughout spacecraft or extraterrestrial habitats is critical. It ensures comprehensive monitoring and allows for redundancy in case of sensor failure. The overlapping capabilities of different sensors can provide a more reliable dataset, enhancing the mission's overall safety and success rate.

The integration of chemical sensors into both robotic and human space missions involves comprehensive planning around design, implementation, and operational strategies. These sensors are fundamental in collecting scientific data and ensuring the environmental conditions necessary for the health and longevity of human crews and the operational integrity of robots. The data provided by these sensors not only aid in immediate decision-making but also enrich our understanding of space environments, thereby contributing profoundly to the field of astrochemistry and space exploration at large.

9.9 Challenges in Operating Sensors in Extreme Conditions

Chemical sensors deployed in space missions must perform under extreme conditions that are typically not encountered on Earth. These include severe temperature variations, high radiation levels, microgravity environments, and vacuum conditions, all of which can

9.9. CHALLENGES IN OPERATING SENSORS IN EXTREME CONDITIONS

significantly affect the performance and durability of these instruments.

Temperature extremes are perhaps the most common challenge. Sensors can be exposed to temperatures as low as -240 degrees Celsius near the lunar poles during the lunar night or as high as 460 degrees Celsius on the surface of Venus. Such conditions require the use of materials and components that retain operational integrity and accuracy. For example, traditional semiconductor-based sensors often fail in these extreme conditions due to semiconductor mobility drastically decreasing at low temperatures or material degradation at high temperatures. Advanced materials such as silicon carbide, gallium nitride, and certain ceramics that have much higher thermal stability are now being used to overcome these limitations.

Radiation encountered in space can induce a variety of harmful effects on chemical sensors, including displacement damage, ionization, and lattice structural changes. Ionizing radiation can cause permanent damage to the sensor's electronic circuits and sensor interface, leading to erroneous readings or complete failure. For instance, sensors used on Mars rovers must contend with both solar and cosmic radiation. Developing radiation-hardened electronics or implementing adequate shielding, while maintaining the sensor's sensitivity and weight effectiveness, is a critical ongoing research and engineering focus.

Microgravity also presents unique operational challenges, primarily affecting fluid-based analytical techniques such as gas chromatography. In microgravity, fluids do not behave as on Earth; for example, buoyancy-driven convection, which assists in heat transfer and mixing processes under normal gravity, is absent. This requires rethinking of the fluid management systems in instruments to prevent bubble formation and ensure proper fluid flow and sample introduction mechanisms. Space mission instruments like the Sample Analysis at Mars (SAM) onboard Curiosity rover have successfully incorporated novel fluidic systems designed for these environments.

Vacuum conditions in space can affect the performance of sensors that rely on conductive measurements or those that are sensitive to air composition and pressure. Outgassing of materials can contaminate sensitive surfaces and degrade sensor performance over time. Vacuum-compatible materials and hermetic sealing are strategies to mitigate these effects, ensuring sensor longevity and reliability.

Moreover, the combination of these factors can lead to synergistic effects. For example, the combination of low temperatures and radiation can result in unexpected material brittleness, which must be accounted for in the design phase of the sensor system.

Due to these extreme conditions, extensive pre-mission testing is paramount. Sensors are subjected to thermal-vacuum chambers to simulate space conditions, radiation testing facilities to simulate cosmic and solar radiation, and drop towers or parabolic flights to test components under reduced or microgravity conditions.

Maintaining consistent sensor performance while overcoming these challenges is crucial for the success of space missions. Continuous advancements in material science, coupled with innovative engineering solutions and rigorous testing regimes, pave the way for developing robust sensors capable of operating in these extreme environments.

Problem Set: 1. Discuss the impact of radiation on semiconductor-based sensors and propose a material that could improve the sensor's resistance to high levels of cosmic radiation. 2. Explain how microgravity affects the operation of chromatographic systems and how the Sample Analysis at Mars (SAM) instrument addresses these challenges. 3. Design a theoretical model for a temperature control system that could maintain a chemical sensor's operational range in extreme thermal environments, like those found on Mercury. 4. Assess the potential effects of outgassing on a vacuum-based spectroscopic sensor and suggest materials that could mitigate this issue.

This section provides detailed insights into the challenges faced by chemical sensors in outer space, emphasizing the necessity for innovative solutions and robust testing to ensure successful space mission outcomes.

9.10 Future Innovations in Chemical Sensing for Space Exploration

As space exploration continues to reach new frontiers, the development of innovative chemical sensors becomes increasingly crucial. The future of chemical sensing in space exploration is poised for transformative advancements that promise to significantly enhance our ability to explore and understand the cosmos. This section will explore key areas of potential innovation including miniaturization, enhanced sensitivity and selectivity, integration of sensor arrays, advances in computational methods, and the evolution of autonomous systems.

Miniaturization remains a central focus in sensor technology development. Reducing the size and mass of chemical sensors without compromising their efficiency is critical for space missions where payload constraints are stringent. Innovative materials such as graphene, carbon nanotubes, and conductive polymers are being explored for their potential to construct nanoscale sensors. These materials not only aid in the miniaturization but also enhance the sensitivity of chemical sensors to detect low-abundance extraterrestrial molecules.

The push for higher sensitivity and selectivity is driving the incorporation of advanced materials and novel transduction mechanisms. For example, molecularly imprinted polymers (MIPs) offer tailored recognition sites for specific molecules, increasing selectivity and sensitivity. Such advancements are essential for the detection of complex organic molecules or potential biosignatures in extraterrestrial environments. Photonic crystal structures and plasmonic materials are also being investigated for their ability to enhance signal outputs by manipulating light at nano dimensions, which could dramatically improve the detection limits of optical chemical sensors.

In addition to individual sensors, the integration of sensor arrays, or 'electronic noses', that can provide a comprehensive analysis of a chemical environment, is gaining traction. These arrays mimic the human olfactory system, providing a holistic view of the chemical landscape on planets like Mars. The data obtained from these sensors can be fed into machine learning algorithms to classify and predict unknown compounds based on learned chemical patterns. This approach can significantly reduce the time needed for data analysis and increase the efficiency of decision-making processes on space missions.

The integration of advanced computational methods is set to revolutionize the way chemical data is processed in space. Machine learning and artificial intelligence (AI) are essential in handling the large volumes of data generated by sensors, providing real-time insights and enabling autonomous decision-making in remote space environments. These computational techniques can predict sensor failures, adapt operational parameters based on environmental data, and efficiently manage the power and resources available to onboard sensor systems.

Lastly, the development of fully autonomous sensing systems is a critical frontier. These systems would not only perform routine analyses but also make proactive decisions about where and when to deploy certain sensors or when to change analytical tactics. The incorporation of autonomous mobility platforms equipped with an array of chemical sensors could drastically enhance the exploration capabilities of rovers and probes.

The future of chemical sensing in space exploration is marked by the convergence of nanotechnology, advanced materials science, artificial intelligence, and autonomous system technologies. These innovations promise to expand our capabilities in space, enabling more thorough exploration, study, and potentially the colonization of extraterrestrial environments. As these technologies evolve, they will not only push the boundaries of space exploration but also ensure that the sensors employed are as efficient, robust, and informative as possible.

Chapter 10

Future Directions in Space Exploration Chemistry

This chapter explores the emerging trends and potential advancements in the field of space exploration chemistry. It discusses the current research frontiers and technological innovations that could revolutionize how we utilize chemical science in space, including advanced propulsion technologies, sustainable life support systems, and the development of new materials designed for the extreme conditions of space. The text evaluates the implications of these advancements for future missions and the potential for new discoveries, emphasizing interdisciplinary approaches and international collaborations to tackle the challenges of space exploration.

10.1 Challenges and Opportunities in Space Exploration Chemistry

The relentless pursuit of space exploration presents a unique amalgam of challenges and opportunities for the field of chemistry. Each step towards understanding and inhabiting extraterrestrial environments demands innovative solutions to unprecedented problems, leveraging the vast potential of chemical science. This section delves into the primary issues and prospects that space exploration poses to chemists and the broader scientific community.

Extreme Environmental Conditions: One of the foremost challenges is the adaptation of chemical processes to extreme conditions found in space—extreme temperatures, vacuum pressures, and radiation levels. Traditional chemical reactions and processes optimized for Earth's conditions may behave unpredictably under off-world conditions. For instance, the low pressure on Mars affects the boiling point of liquids, crucial for designing feasible life support and propulsion systems. Research into materials such as ionic liquids, which have stable liquid ranges under Martian conditions, represents a pivotal adaptation of chemistry to space environments.

Resource Scarcity and Sustainability: The limited availability of resources in space environments compels the need for closed-loop systems, where every output is efficiently recycled back into inputs. Water recovery and oxygen regeneration are areas where chemistry plays a critical role. The Sabatier reaction, which produces methane and water from carbon dioxide and hydrogen, is an excellent example of a chemical process that has been adapted for use on the International Space Station (ISS) for life support.

Energy Efficiency and Propulsion: Developing energy-efficient propulsion systems that can be sustained over long durations in space is another significant challenge. Chemical propulsion, while currently predominant, offers limited efficiency. The development of electrochemical propulsion systems, like those using ionic liquids as propellants, provides an exciting opportunity to enhance thrust efficiency while reducing fuel requirements.

Health Risks from Extended Space Travel: Extended periods of space travel expose astronauts to health risks from cosmic radiation and microgravity, necessitating the development of new pharmaceuticals and nutritional chemistry optimized for space conditions. Building on advanced knowledge of biochemistry and medicine, space pharmacology must devise solutions that are stable, effective, and quick-acting to address the unique medical challenges faced in space.

Analytical Techniques for Unknown Environments: The unknown aspects of extraterrestrial environments require robust, versatile, and precise analytical techniques to conduct in-situ analyses of soils, atmospheres, and other materials. Development of miniaturized, automated, and radiation-resistant analytical instruments is crucial. Techniques such as spectroscopy and chromatography are being adapted to operate in space, offering precise data to support interplanetary explorations.

Interdisciplinary Collaboration and Innovation: The multifaceted nature of space exploration chemistry demands a collaborative approach, intertwining physics, biology, engineering, and environmental science with chemistry. International collaboration not only pools together a vast array of resources and knowledge but also fosters innovative solutions that might not emerge within the silo of a single discipline or nation.

Economic and Ethical Considerations: Beyond the scientific and technical challenges, economic and ethical considerations play a critical role in space exploration chemistry. The development of technologies that minimize environmental impact and are economically viable is crucial. Moreover, as plans for utilizing space-based resources and interplanetary trade evolve, ethical considerations must guide the sustainable and equitable use of extraterrestrial environments.

While the chemistry of space exploration presents formidable challenges, it is rife with opportunities to expand the frontiers of science. Each challenge invites innovative solutions that not only push the boundaries of chemical research but also address fundamental questions about sustainability, collaboration, and our place in the universe. These endeavors not only propel humankind further into the cosmos but also reflect back, offering improvements for life on Earth.

10.2 Advanced Propulsion Systems and Fuel Chemistry

Advanced propulsion systems are crucial for the enhancement of space travel efficiency and the exploration of distant celestial bodies. The chemical underpinnings of fuel production and utilization in space exploration not only dictate the mission's viability but also pose unique technological and environmental challenges. In this context, propulsion technologies such as chemical rockets, electric propulsion, and nuclear thermal propulsion emerge as critical areas of study.

Chemical propulsion remains the backbone of spacecraft launch and maneuver systems, utilizing the combustion of chemical propellants like liquid hydrogen and oxygen to produce thrust. The reaction between these propellants yields water as a byproduct, characterizing a relatively clean emission profile. However, the efficiency of chemical rockets is inherently limited by their specific impulse, urging the need for more advanced propulsion methods.

Electric propulsion systems, including ion and Hall-effect thrusters, present a significant leap towards higher efficiency. These systems utilize electric fields to accelerate ions to high speeds, thereby generating thrust. The primary advantage of electric propulsion is its higher specific impulse compared to conventional chemical rockets. For instance, xenon is often used as a propellant in ion thrusters due to its inert nature and high atomic mass, which provides a favorable balance between ionization energy and thrust output. The physics governing the ionization process and the subsequent acceleration are crucial for optimizing performance and require innovative approaches in electrical engineering and materials science to handle the high-energy environments inside thrusters.

Transitioning to nuclear thermal propulsion offers another promising avenue with potentials to drastically shorten interplanetary travel times. This technology leverages nuclear reactions to heat a propellant like hydrogen, which then expands and ejects out to produce thrust. The chemistry of handling, storing, and reacting nuclear materials safely in the vacuum of space is complex and requires rigorous safeguards against radiation. Moreover, the reaction control and materials that can withstand extreme thermal and radiative conditions involve broad investigations into high-temperature chemistry and radiation chemistry.

Each propulsion method comes with its unique set of fuel chemistry challenges. For chemical propulsion, the focus lies on maximizing the energy density of propellants while ensuring stability and safety under various environmental conditions. In electric propulsion, the ionization efficiency of potential propellants becomes a central concern, necessitating detailed studies into their electronic structure and bond energies. Meanwhile, nuclear thermal propulsion demands innovations in the synthesis and manipulation of nuclear fuel materials that are both efficient and safe.

The development of advanced propulsion technologies also hinges on the synthesis of new materials to handle the high-energy and low-gravity conditions of space. These include alloys and composites capable of withstanding high temperatures, corrosive environments, and radiation without significant degradation.

Incorporating these disparate factors into the design and operation of space missions necessitates a multidisciplinary approach, drawing from physical chemistry, materials science, aerospace engineering, and nuclear physics. The ongoing evolution of propulsion technologies highlights the interdependent progression between chemical innovation and space exploration capabilities.

Problem Set

- Determine the specific impulse of a chemical propulsion system given the following parameters: mass flow rate of $10\,\text{kg}\,\text{s}^{-1}$, exit velocity of $3000\,\text{m}\,\text{s}^{-1}$.

- Discuss the trade-offs between using liquid hydrogen/oxygen propellants and xenon in terms of both performance and environmental impact for missions to Mars.

- Design a hypothetical electric propulsion system, outlining the choice of propellant, design of the ionization chamber, and methods of ion acceleration, given a target specific impulse of $5000\,\text{s}$.

- Analyze the potential challenges of handling and storing nuclear materials for propulsion in the context of a voyage to Jupiter, considering both chemical and radiological safety factors.

- Propose a novel composite material that could be used in the combustion chamber of a nuclear thermal rocket, describing its composition, expected properties, and resistance to high temperatures and radiation.

This problem set serves to integrate theoretical knowledge into practical applications, underlining the comprehensive understanding required for advancing propulsion technologies in space exploration.

10.3 Biochemical Solutions for Long-Duration Space Travel

As future missions aim to venture farther into the cosmos, especially to destinations such as Mars or even the outer planets, the duration of spaceflights will significantly increase. This extension poses pronounced challenges not only in terms of propulsion and materials but crucially in maintaining astronaut health through biochemically enhanced life support systems. This section elucidates innovative biochemical solutions essential for sustaining human life during prolonged space missions. It covers nutrient synthesis, waste recycling, and the biochemical management of closed environments, integrating real-life scenarios and offering mathematical models to deepen understanding.

Long-duration space travel necessitates a reevaluation of traditional life support systems, which must now operate efficiently for extended periods without resupply. One critical component is the development of a bio-regenerative life support system (BLSS). These systems utilize a combination of biological processes and engineering controls to recycle waste, produce food, and maintain air and water quality. For instance, photosynthetic organisms such as algae can regenerate oxygen while providing nutritional supplements. An example is the usage of *Chlorella* algae, which has been demonstrated to efficiently produce oxygen and biomass from carbon dioxide and water under controlled light conditions.

Another focal area in biochemical solutions is the synthesis of essential nutrients in situ. Space travelers cannot carry large reserves of vitamins and other micronutrients due to mass constraints, necessitating novel methods of nutrient synthesis. Bioengineered bacteria that can synthesize vitamin B12, a critical nutrient that is cumbersome to store and transport, represent a potential solution. Employing genetically modified microorganisms that can convert available substrates into complex vitamins could significantly reduce the payload and enhance the sustainability of long-term missions.

Hybrid systems employing both physicochemical and biological processes are being developed to manage waste and produce resources. One methodology that has shown promise is the use of biofilm reactors, which can process organic waste materials into water, gases, and other useful byproducts through microbial activity. For instance, a *Pseudomonas* species might be employed to break down urine into nitrogen, water, and trace gases, with the water being recycled back into the spacecraft's life support system.

The integration of these biochemical systems into the confined and isolated environment of a spacecraft also necessitates rigorous attention to safety and reliability. Mathematical modeling of biochemical

10.3. BIOCHEMICAL SOLUTIONS FOR LONG-DURATION SPACE TRAVEL

cycles within these closed systems is essential to predict and control environmental variables, ensuring the health and safety of the crew. Simulations that model the interactions between human metabolic processes and microorganism activities can help optimize conditions for both human life and microbial processes that support life support system functionality.

To illustrate, consider a scenario where *Rhodospirillum rubrum* is utilized aboard a spacecraft to recycle carbon dioxide. A mathematical model might represent this process, denoting the rate of photosynthesis under varying light conditions, the organism's growth rate, and the subsequent impact on carbon dioxide and oxygen levels. Such models assist in designing robust systems by providing crucial insights into the dynamics of biological processes under spaceflight conditions.

In preparing for these long-duration missions, biochemical research has also seen advancements in the synthesis of pharmaceuticals in space, leveraging microgravity to improve the efficiency and yield of certain chemical reactions. This is particularly relevant as the capability to synthesize medications on-demand could be vitally important given the impossibility of predicting all medical needs over years-long missions.

Finally, while the advancement of biochemical technologies for space bears great promise, these innovations also bring challenges such as containment of genetically modified organisms, prevention of contamination, and ensuring system redundancies. Developing solutions that are resilient to the rigors of space travel, adaptable to unforeseen challenges, and capable of supporting human life for extended periods will be pivotal.

The integration of advanced biochemical strategies into space exploration missions promises not only to enhance the feasibility of long-duration missions but also to bring forward technologies that could benefit sustainable practices on Earth. As we advance, the tailored adaptation and rigorous testing of these systems under simulated and real space conditions will be critical, ensuring that they can support life efficiently and safely wherever humanity's interplanetary ambitions lead.

10.4 Smart Materials for Adaptive Space Structures

The evolution of materials science, particularly in the realm of smart materials, has heralded a new era in the design and implementation of space structures. These materials, characterized by their ability to respond dynamically to environmental stimuli, are pivotal in the development of adaptive structures crucial for the harsh and unpredictable environment of space. This section delves into the composition, mechanisms, applications, and future prospects of smart materials designed specifically for this purpose.

Smart materials for space applications generally fall into several categories based on their responsive nature. Shape memory alloys (SMAs) and shape memory polymers (SMPs) are prominent examples, capable of returning to a pre-defined shape when subjected to a certain stimulus, such as temperature or electrical current. Piezoelectric materials, which generate voltage in response to mechanical stress, and magnetostrictive materials, which change their shape or dimensions under the influence of a magnetic field, are also crucial. Furthermore, electrochromic materials, which change color or transparency in response to voltage changes, offer potential for regulating solar radiation and protecting against space radiation.

The integration of these smart materials into space structures offers diverse functionalities. SMAs and SMPs are especially beneficial in the deployment mechanisms of satellites and space probes. Their capacity to alter shape effectively without external mechanical intervention simplifies the deployment and reduces the risks and costs associated with mechanical systems. Piezoelectric materials are being utilized in energy harvesting devices to convert mechanical stress, such as vibrations from spacecraft machinery, into usable electrical energy. This not only augments the energy supply but also enhances the self-sustainability of space missions. Magnetostrictive materials find their use in precision control systems, where their deformations can be finely tuned for accurate adjustments in the spacecraft's orientation and trajectory. Electrochromic materials are incorporated into windows of spacecraft habitats, where their light modulation properties help in managing the internal light and temperature, contributing to the comfort and well-being of astronauts.

Moreover, the development of smart materials for space structures

10.4. SMART MATERIALS FOR ADAPTIVE SPACE STRUCTURES

does not only focus on functionality but also embraces sustainability. The harsh conditions of space impose severe degradation risks on materials—from extreme temperature fluctuations to high radiation levels. Smart materials are being engineered to possess self-healing capabilities, where microcapsules filled with healing agents are embedded within the material matrix. Upon damage, these capsules rupture and release the healing agent, thereby restoring the material's integrity. This self-healing property significantly extends the life span of space structures, reducing the need for frequent repairs or replacements.

The future of smart materials in space structures looks toward even more integrated and autonomous applications. Current research is focused on developing materials with multiple responsive properties, such as SMPs that also exhibit piezoelectric characteristics. The integration of artificial intelligence and machine learning algorithms with these materials is envisaged to lead to fully adaptive space structures. These structures will not only respond to current environmental conditions but will also be capable of predicting and adapting to future environmental changes, thereby optimizing their performance and safety without human intervention.

The implementation of smart materials in space exploration represents a synergistic confluence of chemistry, materials science, and aerospace engineering. As research progresses, these advanced materials will undoubtedly play a crucial role in shaping the architecture of future space missions, making them more resilient, efficient, and sustainable. The continuous innovation and interdisciplinary research in this field are essential for the realization of ambitious space exploration goals, paving the way for prolonged human presence in space and interplanetary exploration.

The utilization of smart materials in the development of adaptive space structures offers significant advantages by enhancing the functionality, durability, and autonomy of spacecraft and other space-based structures. As we expand our horizons in space exploration, the evolution of these materials will continue to be a cornerstone of technological advancements that will drive future missions. Each step forward in this domain not only promises enhanced mission capabilities but also bolsters the sustainability and safety of space travel, reflecting a profound connection between material innovation and exploratory success. To ensure students grasp these intricate concepts, the following problems are proposed:

CHAPTER 10. FUTURE DIRECTIONS IN SPACE EXPLORATION CHEMISTRY

Problem 1: Describe how shape memory polymers might be used in the development of a mars rover's deployment system. Specify the stimuli responsible for the activation of SMPs in this scenario.

Problem 2: Evaluate the potential advantages of incorporating piezoelectric materials in a spacecraft's energy systems. Explain how these materials could transform mechanical stress into electrical energy during a lunar landing.

Problem 3: Design a conceptual model of a spacecraft window using electrochromic materials. Discuss how such windows could regulate both light and temperature inside the spacecraft, thereby contributing to life support systems.

10.5 Enhanced Analytical Techniques for Space Sample Analysis

As space missions extend beyond the confines of Earth's atmosphere and gravitate towards other celestial bodies, the need for sophisticated analytical techniques becomes paramount. The inherent challenges posed by space environments necessitate the development of robust, sensitive, and versatile analytical tools. This section delves into the advancements in analytical methods designed for analyzing chemical and physical properties of space-obtained samples under the often harsh conditions of extraterrestrial environments.

One of the foremost advancements in this realm is the miniaturization of instrumentation. Traditional Earth-based laboratories are equipped with large, sensitive instruments which are impractical for space due to their size, power consumption, and susceptibility to mechanical failures under extreme conditions. In response, scientists and engineers have developed microanalytical devices that are not only compact but also significantly resilient against the vibrations of launch and the vacuum of space. These include miniaturized spectrometers and chromatographs, which are crucial for molecular and elemental analysis. For instance, microfabricated gas chromatography systems have been successfully implemented in Mars missions to analyze atmospheric samples and volatile compounds in the Martian soil.

Coupled with miniaturization is the integration of autonomous technologies. Space missions can benefit greatly from analytical instru-

10.5. ENHANCED ANALYTICAL TECHNIQUES FOR SPACE SAMPLE ANALYSIS

ments that perform tests without human intervention. Advances in robotics and artificial intelligence have led to the development of autonomous laboratories, capable of conducting complex analytical protocols such as sample preparation, reagent handling, and data interpretation. The Sample Analysis at Mars (SAM) suite aboard the Mars Curiosity Rover exemplifies this, utilizing a combination of chromatography, mass spectrometry, and tunable laser spectroscopy to autonomously analyze soil and atmospheric samples.

Another significant advancement pertains to the durability and recalibration capabilities of space-bound analytical instruments. The exposure to cosmic rays, extreme temperatures, and microgravity can degrade the performance of sensitive instruments. Innovative materials and shielding techniques have been developed to enhance their longevity and reliability. Furthermore, calibration methods that can be conducted in-situ are being integrated to ensure ongoing accuracy of the scientific data collected.

The application of these enhanced analytical techniques is not just limited to understanding extraterrestrial geology or searching for signs of life. They are also critical in monitoring the health of astronauts by analyzing cabin air quality and water purity. For instance, spectroscopic techniques are employed to detect potential contaminants and ensure that life support systems are functioning optimally.

The integration of nanotechnology has offered another leap forward in space sample analysis. Nanosensors, for instance, provide highly responsive and selective detection capabilities, which are vital for the identification of trace amounts of organic compounds or possible toxins. These sensors are engineered to operate effectively in the reduced gravity of space and require minimal power, making them ideal for long-duration missions.

The field of space exploration continually drives the evolution of analytical techniques. By overcoming the unique challenges posed by space environments, these innovative methodologies not only enhance the scientific return from space missions but also ensure the safety and success of these ventures. The continuous refinement and adaptation of these techniques will undoubtedly play a crucial role in future exploratory missions, potentially revealing new insights into the cosmos and furthering our understanding of the universe.

Problems for Consideration

- Design a concept for a miniaturized spectrometer suitable for use on a lunar lander. Consider limitations related to size, weight, and power consumption.

- Propose an autonomous analytical protocol to detect and analyze amino acids on the surface of Europa. Include considerations for sample collection, contamination prevention, and data analysis.

- Evaluate the potential impacts of cosmic radiation on spectroscopic measurements in space. Suggest methods to mitigate these effects.

- Generate a hypothetical calibration procedure for a microfabricated gas chromatography system used on Mars. Discuss how in-situ recalibration could be achieved. *Develop a predictive maintenance schedule for an autonomous laboratory on a Mars rover. Consider factors like instrument usage frequency and environmental impact.*

10.6 Artificial Intelligence in Chemical Research and Monitoring

With the expansion of space exploration, the role of artificial intelligence (AI) in supporting chemical research and monitoring is becoming increasingly critical. AI, primarily through machine learning algorithms and neural networks, has the capacity to analyze complex chemical data sets, predict molecular behavior under various conditions, and optimize chemical reactions, which are crucial in the constrained environment of space missions.

The integration of AI into chemical monitoring systems on spacecraft and extraterrestrial habitats transforms the approach to maintaining air and water quality, ensuring these essential resources meet safety standards for astronaut health. AI systems continuously monitor the levels of various chemical substances, using sensors that detect gases like carbon dioxide, ammonia, and volatile organic compounds. These AI systems learn from the data they collect, enabling them to predict future changes in environmental conditions and alert engineers or astronauts to potential life-supporting failures before they occur.

In propulsion technology, AI plays a pivotal role in the formulation

10.6. ARTIFICIAL INTELLIGENCE IN CHEMICAL RESEARCH AND MONITORING

and testing of new rocket fuels. Through computational chemistry, AI algorithms can simulate the molecular structure of potential fuel candidates and predict their burn efficiency and stability in the harsh conditions of space. This not only speeds up the development of more efficient propulsion systems but also reduces the risk and cost associated with physical experimentation.

Another application of AI in space chemistry is in the synthesis of pharmaceuticals. Given the limited ability to transport large quantities of medications, AI-driven systems equipped with digital synthesis platforms enable the on-demand production of drugs. These systems use AI to analyze medical data and synthesize the required pharmaceuticals with minimal human input, ensuring astronauts receive personalized medication tailored to their physiological conditions during long-duration missions.

AI also enhances the capability of spectroscopic instruments used in the chemical analysis of planetary surfaces and atmospheres. By integrating AI with these instruments, the analytical processes can adaptively focus on detecting elements or compounds of potential scientific interest, improving the efficiency and accuracy of extraterrestrial geochemical research.

In addition to its applications, the incorporation of AI into chemical research and monitoring in space exploration brings several challenges. The reliability of AI systems in extreme space conditions must be thoroughly vetified; their ability to withstand large variations in temperature, radiation levels, and mechanical stress must be assured. Additionally, the decision-making process of AI systems must be transparent and interpretable to ensure that the conclusions and actions recommended by AI are justifiable and understandable by human operators.

To foster these advancements, AI tools are being trained using vast databases of chemical properties and reaction mechanisms both terrestrial and extrapolated for extraterrestrial environments. This training involves not only classic machine learning approaches but also newer, more sophisticated methods like reinforcement learning and deep learning, which can model more complex patterns and adapt more flexibly to new environments.

AI is revolutionizing the field of chemical research and monitoring in space exploration. By enhancing the efficiency, accuracy, and safety of chemical processes critical to space missions, AI not only helps to sustain astronaut lives but also broadens the horizons of what can be achieved in space exploration. As we continue to refine these technologies and integrate them into our space endeavors, the synergy between AI and chemistry is set to play a pivotal role in shaping the future of humanity's journey into the cosmos.

10.7 Space Pharmacology: Chemistry for Astronaut Health

Space pharmacology, a vital frontier for chemical science, extends its applications to the unique and challenging environment of outer space, focusing particularly on maintaining astronaut health under unprecedented conditions. The isolation, radiation levels, microgravity, and other physical constraints in space necessitate an innovative approach to drug development, stability studies, and administration methods.

In the confined and isolated habitats of spacecraft and extraterrestrial bases, astronauts are exposed to conditions that can precipitate various physiological and psychological changes. These include alterations in cardiovascular function, bone density loss, immune system weakening, and enhanced radiation exposure–each providing a unique challenge for pharmacological intervention. Thus, space pharmacology not only needs to adapt existing medications but also develop new ones that are effective under these altered physiological conditions.

One of the major challenges faced in space medicine is the altered pharmacokinetics and pharmacodynamics in astronauts. Drugs behave differently in space due to changes in gastrointestinal absorption, altered plasma drug concentration profiles, and a potentially modified metabolic rate. Research has indicated significant variances in absorption rates, distribution volumes, and elimination processes in microgravity compared to Earth conditions. For instance, the cardiovascular system's response to drugs can deviate in microgravity, necessitating a reevaluation of dosages and delivery mechanisms.

Addressing these issues, recent advancements have explored the development of novel drug delivery systems tailored for space conditions. Nano-encapsulation, a technique that involves enclosing drugs in nanoparticles, helps to prevent degradation and allows for a more controlled release. This is particularly crucial for managing chronic conditions such as osteoporosis or for the delivery of vaccines

10.7. SPACE PHARMACOLOGY: CHEMISTRY FOR ASTRONAUT HEALTH

that must remain stable over long periods. Moreover, 3D-printing technology presents another promising solution, whereby medications can be produced on-demand aboard spacecraft, potentially easing the logistical constraints of carrying a large and diverse drug cache.

Another key area of focus in space pharmacology is the radiation-induced health issues. Radiation poses a severe health risk, increasing the likelihood of cancer and damaging the central nervous system. Here, chemistry steps in to design and synthesize drugs that can mitigate the effects of radiation. Radioprotective compounds that can be administered before exposure to protect astronauts are under rigorous investigation. These compounds function by scavenging harmful radicals produced by radiation or by enhancing the DNA repair mechanisms.

Furthermore, the psychological health of astronauts cannot be underestimated, as prolonged space travel induces various degrees of psychological stress and related disorders. Medications that can effectively address mental health issues without causing adverse side effects in the space environment are crucial. The development of fast-acting anxiolytic or antidepressant formulations that are tailored to avoid interaction with other medications and the unique absorption profiles in space demonstrates the integrative approach needed in space pharmacology.

To ensure the safety and efficacy of medications for space missions, rigorous testing methodologies must be developed. These include advanced simulation environments that mimic space conditions on Earth, allowing for thorough pre-mission testing of drug stability and efficacy. Additionally, collaborations with international space agencies and interdisciplinary partnerships expand the resource network, fostering a more holistic development of space-appropriate pharmacological solutions.

In essence, space pharmacology is poised to become a cornerstone of health management in extraterrestrial environments, delivering targeted, efficacious medication regimes optimized for the unique challenges posed by space travel. The continual evolution of this discipline will significantly dictate the success of long-term manned missions, making it an essential area of study and development within the broader context of space exploration chemistry.

10.8 Eco-friendly and Sustainable Space Exploration Practices

The evolution of space exploration has transcended the boundaries of technological achievements to encompass the pivotal principles of environmental responsibility and sustainability. As space agencies and private enterprises intensify their exploratory efforts, there arises an uncompromising need to integrate eco-friendly practices and sustainability in all facets of interstellar activities. This necessity not only reflects our commitment to preserving outer space but is also crucial for the perpetual sustainability of space missions.

Central to eco-friendly space exploration is the development and deployment of technologies that minimize environmental footprints both in space and on terrestrial support systems. These include the advancement of propulsion systems that utilize non-toxic, less harmful propellants. Traditional chemical propulsion methods rely predominantly on hydrazine, a highly toxic compound that poses significant environmental and health risks during manufacturing, handling, and accidental release. Research is now advancing towards more benign alternatives such as ionic liquids or green propellants like hydroxylammonium nitrate fuel/oxidizer mixture (AF-M315E), which promises a reduction in toxicity and carcinogenic risks, contributes to lower greenhouse gas emissions, and increases fuel efficiency.

Another critical aspect of sustainable space exploration involves the architecture of spacecraft and their support systems. Here, the emphasis is placed on the use of recyclable materials and modular designs that allow for the reuse and repurposing of components. This approach not only reduces the waste generated during spacecraft manufacturing but also extends the operational lifespan of spacecraft, effectively minimizing resource expenditure. Moreover, the modular nature facilitates easier upgrades and maintenance, adapting to new technological advancements without necessitating the construction of entirely new structures.

The recycling and refining of resources on celestial bodies, known as in-situ resource utilization (ISRU), presents another sustainable approach. ISRU can significantly decrease the amount of material launched from Earth, thus reducing payload weights and associated costs and environmental impacts. Techniques such as the extraction

10.8. ECO-FRIENDLY AND SUSTAINABLE SPACE EXPLORATION PRACTICES

of water from lunar regolith or the production of fuel from Martian atmospheric CO_2 are examples of how space missions can become self-sustaining, thereby reducing the reliance on Earth-based resources.

Life support systems in space crafts and extraterrestrial bases pose significant challenges in terms of sustainable chemistry. Here, innovation revolves around closed-loop systems capable of recycling air and water continuously. Advanced bioregenerative life support systems that integrate microalgal or plant-based components not only recycle CO_2 and waste products but also contribute to food production, thus creating a self-sustaining habitat for astronauts during long-duration missions.

Furthermore, space exploration has to consider the impact of extraterrestrial activities on the unblemished environments of other celestial bodies. The implementation of planetary protection protocols ensures that contamination of both Earth and foreign planets is minimized. This involves meticulous sterilization processes for spacecraft and the development of protocols to reduce the accidental transport of Earth-originating life, preserving the integrity and natural state of celestial bodies for scientific investigation and future colonization.

Analyzing the toxicity and long-term environmental effects of materials and chemicals used in space missions is vital for sustainable practices. This facilitates informed choices about materials and construction methods that reduce ecological footprints and enhance the safety of astronauts. Research into alternative materials and sustainable extraction methods continues to be a central theme in the chemistry of space exploration.

Cultivating eco-friendly and sustainable practices in space exploration is not merely an option but an essential strategy for the future of interplanetary science and technology. With careful consideration and continued innovation, these practices promise to support extensive, long-term human and robotic presence in space while safeguarding both Earth's and extraplanetary environments. Engaging in sustainable methodologies will be instrumental in achieving the dual goals of exploration and environmental stewardship in the vast expanse of space.

CHAPTER 10. FUTURE DIRECTIONS IN SPACE EXPLORATION CHEMISTRY

10.9 Interplanetary Trade and Utilization of Space Resources

The paradigm of interplanetary trade and the utilization of space resources is a pivotal element in the sustainable expansion of human presence across the solar system. The extraction and processing of space-based resources not only promises to reduce the Earth-bound payload requirements but also scales the economic feasibility of continued space exploration. Moreover, this approach can lead to the development of new technologies and processes that cater specifically to the harsh environments of outer space.

Understanding the chemical composition and properties of celestial bodies is foundational to this endeavor. Recent advancements in spectroscopy and remote sensing techniques have allowed for more detailed analyses of asteroids, the Moon, and other planetary bodies, revealing a wealth of valuable materials such as water ice, metals, and minerals. For instance, water extracted from lunar poles or asteroid materials can be split into hydrogen and oxygen, critical components for rocket fuel. This not only supports propulsion but also life support systems, through the production of breathable oxygen and potable water.

The chemical processing of these materials in space necessitates the development of robust technologies that can function under extreme conditions of vacuum, temperature fluctuations, and radiation. Techniques such as solvent extraction and electrochemical processing, which are commonly used on Earth for mineral and metal recovery, must be adapted or entirely reengineered for use in microgravity and other space-specific conditions. The feasibility of space-based processing plants relies heavily on advancements in materials science, particularly in the development of materials that can withstand prolonged exposure to space weathering without significant degradation.

Additionally, the concept of 3D printing has been leveraged to utilize in-situ resources, potentially transforming raw materials directly into structural components for habitats or machinery without the need for transport back to Earth. This could significantly drive down costs and logistical complexities associated with building and maintaining extraterrestrial bases. For example, regolith from the Moon has been proposed as a building material, potentially being processed and printed directly on the lunar surface to construct base structures.

From an economic perspective, the legal and economic frameworks for interplanetary trade are still in their nascent stages. The Outer Space Treaty of 1967, which forms the basis of international space law, does not currently accommodate the private ownership of extraterrestrial property, which poses a significant challenge to the commercial exploitation of space resources. As such, there is a pressing need for new policies that not only regulate but also encourage the ethical and sustainable commercial use of space resources.

Incentivizing international cooperation can also amplify the benefits derived from space resources. Collaborative efforts can lead to the sharing of technological advancements, cost reductions through shared missions, and the establishment of common standards and practices that ensure transparency and environmental responsibility.

The section concludes by examining how the utilization of space resources and the chemistry involved in their processing might evolve. As technology advances and legal frameworks adapt, an era of flourishing interplanetary trade could emerge, fundamentally reshaping how humans interact with space. It is a field rich with challenges but equally abundant with opportunities for future chemical research and technological innovation aimed at sustainable space exploration and utilization.

10.10 Educational and Collaborative Projects in Space Chemistry

As the frontier of space exploration expands, the necessity for educational initiatives and collaborative projects in space chemistry becomes increasingly crucial. These endeavors aim to foster a new generation of chemists equipped with the knowledge and skills to tackle the unique challenges posed by space environments. Educational and collaborative projects in space chemistry not only prepare future scientists but also stimulate innovation and broaden the scope of research and development in this field.

One of the primary goals of educational projects in space chemistry is to integrate advanced scientific knowledge with practical applications relevant to space exploration. For instance, university programs around the globe have begun incorporating specialized courses that focus on the chemistry of space travel, life support systems in spacecraft, and the utilization of extraterrestrial resources. These courses often include hands-on laboratory experiments designed to simulate space conditions, thereby providing students with real-world problems solving experiences in a controlled environment.

Collaborative projects, on the other hand, extend beyond educational institutions. They often involve partnerships between academic institutions, space agencies like NASA and ESA, and private aerospace companies. An exemplary collaborative project is the development of life support systems that recycle water and air aboard spacecraft. This project not only applies principles of chemistry but also integrates biological and engineering solutions, demonstrating the interdisciplinary nature required in space chemistry research.

Space simulation facilities, such as the Mars Simulation Facility (MSF) at various universities, allow researchers and students to test the chemical stability of materials meant to withstand Martian conditions. These facilities provide invaluable data that helps in the formulation of materials that are optimized for use on Mars, through collaborative research efforts involving chemists, material scientists, and planetary geologists.

Moreover, the international aspect of space exploration necessitates global cooperation. The International Space Education Board (ISEB), for instance, is an initiative that promotes an international educational environment. Through programs coordinated by ISEB, students from different countries work together on projects such as the analysis of soil samples from the Moon, using remotely operated rovers. These projects not only help students develop their technical skills but also encourage cross-cultural communication and cooperation, essential skills in the increasingly collaborative field of space exploration.

Furthermore, virtual reality (VR) technology has been leveraged to create immersive educational experiences that expose students to the chemical elements of space travel without leaving their classrooms. VR simulations allow students to conduct experiments in a virtual space station laboratory, offering a unique method to engage with complex chemical concepts actively.

Educational and collaborative projects in space chemistry are pivotal in preparing a skilled workforce capable of advancing the goals of future space missions. Through coursework, simulated environments,

10.10. EDUCATIONAL AND COLLABORATIVE PROJECTS IN SPACE CHEMISTRY

and international cooperation, these projects develop the expertise and innovation necessary to overcome the challenges of space exploration. This dynamic integration of education and collaboration not only enhances the understanding and application of space chemistry but also ensures a robust, interdisciplinary approach to solving the mysteries of the cosmos.

www.ingramcontent.com/pod-product-compliance
Lightning Source LLC
Chambersburg PA
CBHW052151220526
45471CB00004B/1633